Birds of the
Central Platte River Valley
and Adjacent Counties

Mary Bomberger Brown and Paul A. Johnsgard

Abstract

The central Platte River Valley region of Nebraska is described ecologically, and defined as encompassing 11 counties and nearly 10,000 square miles, and extending about 120 miles from the western edge of Lincoln County to the eastern edge of Merrick County. At its center is the Platte River, the historic spring staging area for Sandhill and Whooping cranes, five species of geese, and millions of waterfowl and water-dependent birds, in addition to providing the breeding habitats for more than 100 other bird species. Collectively, at least 373 bird species have been reported from the Central Platte Valley, making it the most species-rich bird location in Nebraska, and of the most species-diverse regions in the Great Plains. The abundance, distribution and habitats of these species are summarized, with special consideration given to the Valley's three nationally threatened and endangered birds, the Whooping Crane, Interior Least Tern, and Piping Plover, and the now probably extinct Eskimo Curlew. Also included are a species checklist, a list of 82 regional birding sites, and a bibliography of 130 citations.

Birds of the
Central Platte River Valley
and Adjacent Counties

Mary Bomberger Brown
and Paul A. Johnsgard

Zea Books
Lincoln, Nebraska
2013

Text and illustrations copyright © 2013
Mary Bomberger Brown and Paul A. Johnsgard.

ISBN 978-1-60962-030-1 paperback
ISBN 978-1-60962-031-8 ebook

Set in Constantia types. Design and composition by Paul Royster.
Zea Books are published by the University of Nebraska–Lincoln Libraries.

Electronic (pdf) edition available online at http://digitalcommons.unl.edu/zeabook/
Print edition can be ordered from http://www.lulu.com/spotlight/unllib

Table of Contents

Table of Contents *(continued)*

List of Figures

List of Figures (*continued*)

Acknowledgments

We thank Linda Brown, Alice Heckman, Les Howard, Allison Johnson, and Joel Jorgensen for useful conversations about the birds of the central Platte River Valley as we prepared the material for this book. We thank the University of Nebraska's School of Natural Resources and School of Biological Sciences, Nebraska Game and Parks Commission, and Nebraska Environmental Trust for their support. We appreciate receiving the approval of *Natural History* magazine (Eskimo Curlew) and *Prairie Fire* newspaper (Sandhill and Whooping cranes) to include previously published material in this book.

Figure 1. Map of the Central Platte Valley, showing the counties covered in this book (outlined), plus adjacent counties. Illustration by Mary Bomberger Brown.

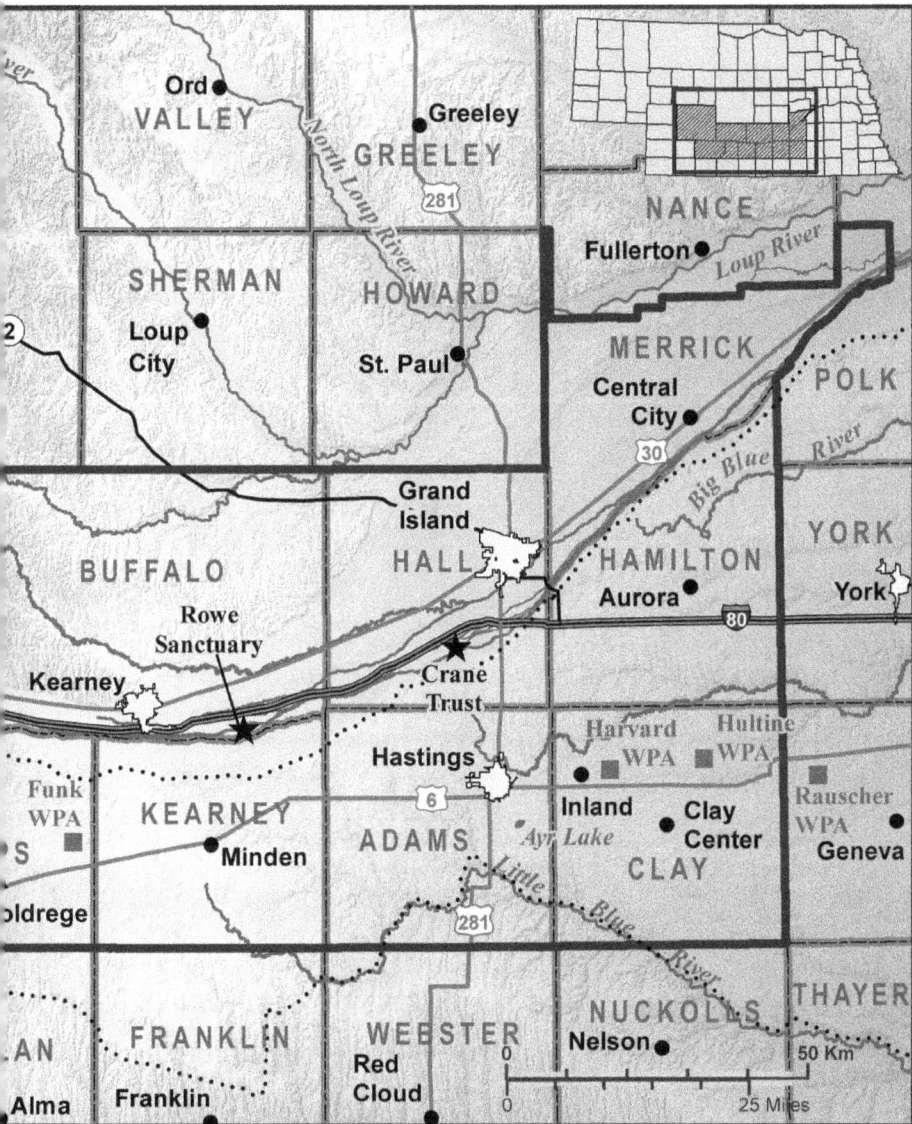

Introduction

Naturalists know Nebraska's central Platte River Valley as the seasonal home to roughly a half-million Sandhill Cranes (*Grus canadensis*). Every spring it also hosts uncountable numbers of waterfowl and shorebirds, as well as threatened or endangered birds such as the Whooping Crane (*Grus americana*), Interior Least Tern (*Sternula antillarum athalassos*) and Piping Plover (*Charadrius melodus*). Besides the cranes, terns and plovers, the central Platte River Valley supports a world-renowned migration of geese and ducks and a remarkable migration of shorebirds, including what is probably the Buff-breasted Sandpiper's (*Tryngites subruficollis*) most important spring staging area between its South American wintering grounds and its arctic breeding grounds. As ecologists are quick to point out, in conserving a few very rare and beautiful species by keeping the river healthy and its associated wetlands intact, we are also protecting the multitude of other living things that are the underpinnings of the entire river ecosystem. Early efforts in protecting the river, especially for cranes, included the establishment of the Rowe Sanctuary and Iain Nicolson Audubon Center (hereafter referred to as Audubon's Rowe Sanctuary) in 1974, and the Platte River Whooping Crane Maintenance Trust (renamed The Crane Trust in 2012), established in 1978.

The Platte River ecosystem must be viewed as a whole, geographically extending across three states and biologically across a multitude of species. This remarkable river, like other Great Plains rivers, has at times been disrupted and nearly destroyed, partly as a result of drought, but largely through human exploitation. Pressured by federal agencies regarding license renewals for Kingsley Dam operations and irrigation permits in relationship to the classification of the central Platte as critical habitats for several threatened or endangered species, the many stakeholders were forced to try to find common ground for saving the river and its resources. Negotiations were initiated in 1997 among a group of diverse interests including federal, state and local agencies, agricultural and irrigation organizations, municipalities, conservation groups and others. Under political pressure and at the request of Congress, the National Research Council of the National Academies of Science performed a study (National Research Council of the National Academies, 2005) that substantiated what biologists had long recognized–that the river is a fragile ecosystem with a complex, interdependent ecology, unbounded by state borders, and a resource in dire need of protection and restoration.

An environmental study by the National Academies of Science (U.S. Department of the Interior-U.S. Fish and Wildlife Service, 2006) examined options for maintaining and improving habitat for cranes, terns, and plovers, as well as an endangered fish, the pallid sturgeon (*Scaphirhynchus albus*). A nearly decade-long period of negotiations among the diverse interested parties was culminated by a 2007 three-state and federal multi-year agreement among Nebraska, Colorado, Wyoming, and the U. S. Department of Interior. This agreement, the Platte River Recovery Implementation Program (hereafter referred to as the PRRIP), set forth a framework for a habitat protection and improvement program for the Whooping Crane, Piping Plover, and Interior Least Tern (Freeman, 2010).

As valuable as these programs and studies are, none can assess adequately the immeasurable esthetic, economic and ecologic values of a living and vibrant Platte River. We may be able to count all of the central Platte River Valley's Sandhill Cranes, Interior Least Terns, and Piping Plovers, but it is harder to evaluate the ecological importance of the uncountable number of species that comprise the river ecosystem, the regional economic benefits resulting from increased ecotourism (Lingle, 1992; Fermata. Inc. 1996; Eubanks et al., 1998; ECONorthwest, 2006), or the unbounded joy and pleasure the region gives to the tens of thousands of people who annually travel to Nebraska in spring to experience it.

This book has been written in the hope that Nebraska residents and visitors to the central Platte River Valley will gain a greater appreciation for the region's amazing spring migration, its ornithological diversity, and its international ecological importance. Their futures depend on the continued survival and vitality of the beautiful Platte River, together with its associated wetlands, prairies, and riparian woodlands. By doing so, we are also preserving an integral part of Nebraska's human history. We know that the Platte River's sandy shorelines were trodden on by some 300,000 people making their way west along the Oregon and Mormon trails during the mid-1800's, and were camped on by the Pawnees and earlier Native Americans who had lived there for thousands of years before. Their long-vanished footprints and scattered artifacts help to remind us that the central Platte River Valley is as significant a national historic site as is Chimney Rock, Scotts Bluff, or the prairie-covered hills above Wounded Knee Creek.

Part I.

The Central Platte Valley

Regional Geography and Recent History

The Platte River gracefully makes its way from west to east across Nebraska like a satin ribbon, descending as it crosses the state at an average rate of about nine feet per mile. Its waters are fed from above by streams emerging from the snow-covered mountains of Colorado and Wyoming, and nourished from below by artesian seepage from the Ogallala aquifer. This aquifer is the repository of the remains of late Pleistocene (10,000–20,000 years ago) mountain glacier meltwaters. As the waters of the aquifer seep eastward from the mountains along a bedrock base, they provide a major source of much of the Platte River's flow. The aquifer is largely centered below and deepest in the Sandhills region, northwest of the central Platte River Valley and has been estimated to hold some two billion acre-feet of water (Johnsgard, 2008; Freeman, 2010). Through spring-fed streams originating in the Sandhills, the aquifer also significantly supplements the flow of the lower Platte, especially eastward from the confluence of the Loup and Platte rivers at Columbus.

As it has flowed across the landscape for uncounted millennia, the Platte River has provided a convenient passageway for plants and animals. Prominent among the river's users have been humans, who probably for ten thousand years or more have drunk from its waters and hunted and fished along its shoreline (Johnsgard, 2005). The Platte River's waters have allowed for the establishments of some of Nebraska's major cities, such as Scottsbluff, North Platte, Kearney and Grand Island, as well as dozens of smaller towns and villages from Morrill to Plattsmouth. Easy access to the aquifer has allowed well-based irrigation to flourish in the Platte Valley, as well as surface-fed irrigation via diversions from Nebraska's reservoirs.

This combination of abundant water and fertile soils has made the central Platte River Valley the geographic heart of Nebraska. The region contains approximately 150 river miles of the Platte Valley, extending from three miles west of Sutherland to four miles east of Silver Creek. It also includes parts of the Wood, South Loup, Big Blue and Little Blue rivers, which, except for the Big Blue and Little Blue rivers, all lay within the greater Platte River drainage. The region's larger impoundments include Sutherland (3,000 acres) and Maloney (1,600 acres) reservoirs in Lincoln County, Medicine Creek (1,700 acres) and Red Willow (1,600 acres) reservoirs in Frontier County, Johnson Lake (2,000 acres) and Elwood Reservoir (1,300 acres) in Gosper County, and Midway (600 acres) and Plum Creek (250 acres) reservoirs in Dawson County. None of these irrigation-related reservoirs has sufficiently stable water levels to provide significant nesting habitat for aquatic birds, but they do provide migratory stopover sites for many, especially waterfowl and gulls. The adjacent Rainwater Basin to the south of the Platte Valley historically supported thousands of wetlands of enormous importance to waterfowl and other aquatic wildlife. The Rainwater Basin currently supports about 84 seasonal wetlands occupying 28,600 acres, including 21,742 federally owned acres, and about 6,900 state-owned acres (www.rwbjv.org; Johnsgard, 2012).

This region receives sufficient annual precipitation so that a dry land farming- and ranch-based economy developed, supplemented in the middle 1900's through irrigation infrastructure development. Later, agricultural technology, especially the development and expansion of center-pivot irrigation, increased use of fertilizers and pesticides, more efficient planting and harvesting techniques and genetically modified crops, have transformed the central Platte Valley into an agricultural powerhouse (Johnsgard, 2008). By 2005 there were over 91,000 registered groundwater wells in the state, or more than one per acre averaged over the state as a whole. Nearly 95 percent of all the groundwater removed from these wells is used for agricultural purposes, with 7.2 million acres being irrigated by groundwater, including over 2.1 million acres in the central Platte Valley alone. All of these changes in have had significant influences on the region's land, vegetation and wildlife (Johnsgard, 2008).

As a result of the 2007 PRRIP agreements, irrigators in the Central Platte valley will be forced to provide some 150,000 additional acre-feet of water to maintain habitat for endangered species by 2019. This stretch of the Platte is already over-appropriated as to irrigation limits, so it is possible that thousands of acres of land now being irrigated by wells or surface waters will need to be returned dry-land farming. Considerations are underway to use water moving through unlined irrigation canals from Lake McConaughy water to be allowed to soak into the ground and recharge the underlying aquifer, which since the advent of surface irrigation has risen substantially, and might support some new well-drilling activities.

The central Platte River Valley region, as defined here, encompasses 11 counties and 9,745 square miles (approximately 2,524,000 hectares) from the western edge of Lincoln County to the eastern edge of Merrick County. The associated counties include Adams, Buffalo, Clay, Dawson, Frontier, Gosper, Hall, Hamilton, Kearney, Lincoln, Merrick and Phelps counties (Fig. 1). This region overlaps substantially with the Rainwater Basin in Adams, Clay, Gosper, Hall, Kearney, and Hamilton counties.

Regional Terrestrial Ecosystems

The native vegetation of the central Platte River Valley was an early casualty of human settlement in the region. However, remnants of the region's pre-settlement native ecosystems do still survive, especially to the north of the region, where it abuts the southern edge of the Nebraska Sandhills.

The Nebraska Sandhills remain a semi-wilderness that is almost entirely covered with native grassland vegetation. This unique community (Sandhills mixed-grass prairie), adapted to the fragile sandy substrate, and flourished after it was formed by the deposition of sand and fine gravels produced by melting glacial melting during late Pleistocene times. The sand was shaped by the prevailing northwest winds into relatively linear dunes, their long axes generally oriented from northeast to southwest. In the central Sandhills, where the sand is often more than 500 feet deep, some dune crests may reach heights of about 300 feet, and the inter-dune valleys are often so close to the underlying aquifer as to produce lush sub-irrigated meadows or shallow, vegetation-rich wetlands (Bleed and Flowerday, 1989; Johnsgard, 1995). Although the dunes are now mostly stabilized with vegetation, during recurring periods of long-term drought over geologic time, they have been set into motion, until they again have become stabilized by a more mesic climate cycle permitting vegetation to re-colonize the region (Loope and Swinehart, 2000).

Closely adjacent to the Platte Valley itself are native grasslands that support what may be the largest surviving population of Greater Prairie-Chickens (*Tympanuchus cupido*) of any state, and an even larger population of Sharp-tailed Grouse (*Tympanuchus phasianellus*). Nebraska is perhaps the easternmost portion of the Great Plains where one can reasonably expect to see such classic grassland symbols as Burrowing Owls (*Athene cunicularia*), Prairie Falcons (*Falco mexicanus*), Golden Eagles (*Aquila chrysaetos*) and Ferruginous Hawks (*Buteo regalis*). Just to the north of the central Platte River Valley are the Nebraska Sandhills, a near-wilderness of approximately 19,000 square miles (49,200 km²) with large populations of such classic grassland birds as the Long-billed Curlew (*Numenius americanus*), Upland Sandpiper (*Bartramia longicauda*), Horned Lark (*Eremophila alpestris*), and a dozen species of grassland sparrows. And, just to the south of the central Platte River Valley is the Rainwater Basin, a landscape of hundreds of spring meltwater ponds and other wetlands that seasonally support a great diversity of water-dependent species. Few other places in North America can provide so much appeal to birders and naturalists.

There are many bird species typical of the Sandhills grasslands, such as the Greater Prairie-Chicken , Sharp-tailed Grouse , Swainson's Hawk (*Buteo swainsoni*), Northern Harrier (*Circus cyaneus*), Loggerhead Shrike (*Lanius ludovicianus*), Horned Lark, Lark Bunting (*Calamospiza melanocorys*), Grasshopper, Lark, and Vesper sparrows (*Ammodramus savannarum, Chondestes grammacus, Pooecetes gramineus*), and Western and Eastern meadowlarks (*Sturnella magna, S. neglecta*) (Labedz, 1989; Johnsgard, 2005). Labedz (1989) established that the Sandhills region supports 23 species of birds that are ecologically associated with grasslands. There is also an associated mammal community of more than 50 species, nearly half of which are rodents and lagomorphs (Freeman, 1989). Except in the wet-meadow valleys, the sandy soil is too loose to allow for larger tunneling rodents such as prairie dogs (*Cynomys* spp., and their many ecological associates), but smaller rodents such pocket mice (*Perognathus* spp.) and kangaroo rats (*Dipodomys* spp.) survive very well, providing important foods for raptors such as Swainson's and Red-tailed hawks (*Buteo jamaicensis*), Northern Harriers, and Short-eared Owls (*Asio flammeus*).

South of the Sandhills and the alluvial plain of the Platte Valley, and extending into Kansas and Oklahoma, lies a vast landscape comprised of wind-blown silty soils, called loess (Farrar, 1997). This loess-based soil provides a fertile base for easily cultivated (and easily eroded) soils. The loess mantle now blanketing the Rainwater Basin evidently came out of the Platte Valley, and was transported as recently as 20,000–30,000 years ago (Kuzila, 1994). In eastern Lincoln County, southwestern Dawson County, and over much of Frontier and Gosper counties this mantle of loess soil is especially thick, forming an region of steep hill-and-canyon topography (Farrar, 1993). The native vegetation of the Loess Hills region is mostly mixed-grass prairie and is a combination of the low-stature perennial grasses and forbs of the short-grass prairie mostly found farther west, plus taller, more moisture-dependent plants shared with the tallgrass prairies of eastern Nebraska (Hopkins, 1949; Branson, 1952; Rothenberger and Bicak, 1993; Rothenberger, 1998). Eastern red cedar (*Juniperus virginiana*) is often a conspicuous part of this diverse vegetational complex, especially on north- and east-facing slopes (Rolfsmeier and Steinauer, 2010). Scattered clusters of ponderosa pines (*Pinus ponderosa*) occur locally in the Sandhills as well as farther west in the North Platte River Valley, their erratic occurrences representing botanic relics of an earlier, much cooler, climate.

The bird life of the loess mixed-grass prairies is similar to that of other mixed-grass prairie regions, with an abundance of seed-eating and ground-nesting birds, most of which are concealingly plumaged and that become conspicuous only when engaged in advertising songs and courtship displays. Typically, mixed-grass and tallgrass prairies often have less than ten species of grass-dependent breeding birds present, although 30 species are especially characteristic of mixed-grass and tallgrass prairies in the Great Plains collectively, and might be considered as grassland endemics. The five breeding species having the highest mean densities in these grasslands are the Western and Eastern Meadowlark, Horned Lark, Dickcissel (*Spiza americana*) and Lark Bunting (Johnsgard, 2001a). All of these species have loud and/or complex songs that are often uttered in flight, when the birds' underparts and tail patterns are effectively exhibited. These species, plus the Upland Sandpiper, are also usually the most abundant summer birds in tallgrass prairies throughout North America (Wiens, 1973; Wiens and Dyer, 1975).

In Sandhills–loess transition areas, as is well developed in southwestern Lincoln County, a distinctive shrub-and-grassland community is often present. It is easily identified by the presence of sandsage, along with yucca and many sand-adapted mid-height grasses typical of the Nebraska Sandhills to the north (Wehrman, 1961; Rolfsmeier and Steinauer, 2010).

At the western edge of the central Platte Valley region, upland soils can support only shortgrass prairie, where drought-tolerant cool-season grasses such as blue grama (*Bouteloua gracilis*) are often important and scattered shrubs or subshrubs are commonly present. Native plant species diversity here is low, with many invasive grasses typical. However, this historic home of bison (*Bison bison*) and pronghorns (*Antilocapra americana)* is still populated with a few reminders of our western heritage, such as remnant colonies of black-tailed prairie dogs (*Cynomys ludovicianus*; Johnsgard, 2005). In Nebraska's far western Panhandle these grasslands are the nesting areas of McCown's Longspur (*Rhynchophanes mccownii*) and, in extremely short-grass and bare-ground combinations, the Mountain Plover (*Charadrius montanus*).

Regional Wetlands

The natural wetlands of the central Platte Valley have largely disappeared, as they were drained and converted to agriculture many years ago. However, in recent years, small wetlands have been excavated or renovated to provide resting areas for spring migrant birds, as do the remaining botanically diverse wet meadows, oxbows and seasonal ponds caused by spring flooding (Nagel, 1981; Nagel and Kolstad, 1987). These wetlands are rich in aquatic organisms used by foraging birds (Whiles and Goldowitz, 2005). They also support a diversity of vertebrates, including over 200 species of birds and at least 15 species of lizards and reptiles (Jones et al., 1981; Geluso and Harner, 2011). The breeding birds of these biologically rich meadows include a few tallgrass specialists that usually are found much farther east, such as the Henslow's Sparrow (*Ammodramus henslowii*) (Kim, 2005; Kim et al., 2008).

At the southern edge of the Nebraska Sandhills, a few of the region's Sandhills wetlands can be seen. These include Lamplaugh Lake, Whitehorse Marsh, and Birdwood Creek, typical spring-fed Sandhills stream habitats with many unspoiled wet meadows. Such spring-fed creeks are warm enough to remain open well into the winter, providing important seasonal roosting and foraging habitat for ducks, geese and Trumpeter Swans (*Cygnus buccinator*). The birds of the Sandhills wetlands are relatively well known (Oberholser and McAtee, 1920; Labedz, 1989; Novacek, 1989; Johnsgard, 2005). Labedz (1989) established that the Sandhills wetlands areas support 137 species of breeding birds. The wetlands provide habitats for breeding populations of many species that are otherwise rare, local or absent elsewhere in the state. Among these are the Cinnamon Teal (*Anas cyanoptera*), Redhead (*Aythya americana*), Trumpeter Swan, Greater Prairie-Chicken, Long-billed Curlew (*Numenius americanus*), Black-necked Stilt (*Himantopus mexicanus*), Black-crowned Night Heron (*Nycticorax nycticorax*), White-faced Ibis (*Plegadis chihi*), American Bittern (*Botaurus lentiginosus*) and Eared Grebe (*Podiceps nigricollis*).

South of the central Platte Valley is a gently rolling region known as the Rainwater Basin. It is a loess-mantled plain, with just enough variation in topography for wetlands to form in low areas where snowmelt or precipitation accumulates. The region is about 1.16 million hectares (4,400 square miles) in area, and extends from eastern Gosper County east to Polk, York and Fillmore counties. Its northern boundaries adjoin the Platte Valley, and its southern limits reach the northern parts of Harlan, Franklin, Nuckolls and Thayer counties (Jorgensen, 2012). The region is often geographically divided into eastern and western Rainwater Basins, with the division line passing though Adams County.

By virtue of the region's easily cultivated soils and nearly level terrain, agriculture in the Rainwater Basin developed rapidly. Many wetlands were drained, and many of those that remained were filled-in by the blowing winds of the Dust Bowl years of the 1930's or by sediments carried in by run-off from nearby agricultural lands. By the 1960's more than 80 percent of the region's wetlands that had been present early in the century were gone, and by the 1980's about half of those remaining had also disappeared (Farrar, 1982, 1996). By the early 2000's over half of the region's land area was planted to corn, and about 20 percent to soybeans. By that same time wet meadows and early-succession wetlands each comprised only 0.7 percent of land-use in the region, late-succession wetlands 0.4 percent, and emergent marshes about 0.1 percent (Bishop et al., 2009; Jorgensen, 2012). LaGrange (2005) estimated that the Rainwater Basin had only 34,103 acres (13,800 hectares) of surviving wetlands by the early 2000s.

Jorgensen (2012) provided a thorough summary of the birds of the entire Rainwater Basin region, and an analysis (2004) of the shorebird migration through the eastern portion of the Rainwater Basin. During five years of spring and three years of fall observations, Jorgensen found 38 species of shorebirds occupying the region. During spring, the greatest numbers seen were White-rumped Sandpiper (*Calidris fuscicollis*), Wilson's Phalarope (*Phalaropus tricolor*), Semipalmated Sandpiper (*Calidris pusilla*), Long-billed Dowitcher (*Limnodromus scolopaceus*), Stilt Sandpiper (*Calidris alpina*), and Baird's Sandpiper (*Calidris bairdii*), while during fall the Pectoral Sandpiper (*Calidris melanotos*) was seen in greatest numbers, followed by Long-billed Dowitcher, Lesser Yellowlegs (*Tringa flavipes*), Least Sandpiper (*Calidris minutilla*), and Stilt Sandpiper. The total bird list for the Rainwater Basin is about 250 species, of which over 120 species are reported as regional breeders; including at least 33 waterfowl species, of which 11 are breeders, and over 50 species of shorebirds and gulls, nearly all of which are migrants (USFWS, 2005). Besides supporting an estimated 7–14 million migrating waterfowl, 200,000–300,000 shorebirds use the Basin wetlands during spring (Farrar, 2004; LaGrange, 2005). Consequently, the Rainwater Basin has been identified by the Nebraska Natural Legacy Project as one of Nebraska's Biologically Unique Landscapes (Schneider et al., 2011) and a Landscape of Hemispheric Importance by the Western Hemisphere Shorebird Reserve Network (http://www.whsrn.org/site-profile/rainwater-basin).

Regional Breeding Birds

In a study of 13 south-central Nebraska counties, Faanes and Lingle (1995) estimated that the central and upper Platte River Valley supported 125 species of breeding birds. Thirty-five species were judged to be the most abundant breeders in the region, and represented 95 percent of the overall breeding community. Grasshopper Sparrows (*Ammodramus savannarum*), Western Meadowlarks, and House Sparrows (*Passer domesticus*) comprised over 26 percent of the total estimated breeding bird population. Common Grackles (*Quiscalus quiscula*) and Red-winged Blackbirds (*Agelaius phoeniceus*) contributed another 13 percent of the total population

Of the 35 most common breeding species, the (Ring-necked Pheasant (*Phasianus colchicus*), House Sparrow and European Starling (*Sturnus vulgaris*) are non-native and to varying degrees these interfere with native species. Many of the region's most common nesting birds are Neotropic migrants, which annually return from wintering areas in Mexico, Central America and South America. Among these are the Western and Eastern kingbirds (*Tyrannus verticalis, T. tyrannus*), Upland Sandpiper (*Bartramia longicauda*), Bobolink (*Dolichonyx oryzivorus*), Baltimore and Bullock's orioles (*Icterus galbula, I. bullockii*), Orchard Oriole (*Icterus spurius*), Common Nighthawk (*Chordeiles minor*), Brown Thrasher (*Toxostoma rufum*), Common Yellowthroat (*Geothlypis trichas*), and Barn Swallow (*Hirundo rustica*). The Chimney Swift (*Chaetura pelagica*), instead of retaining its historical pattern of nesting in the hollows of large trees, is now entirely a city-dweller, attaching its nests to the vertical surfaces of chimneys.

Of the region's 37 most numerous breeding species, the Northern Flicker (*Colaptes auratus*) and Red-headed Woodpecker (*Melanerpes erythrocephalus*) are of special ecological importance. Their tree-excavated nesting cavities provide nest sites for such native cavity-nesters as the House Wren (*Troglodytes aedon*), Tree Swallow (*Tachycineta bicolor*), Black-capped Chickadee (*Poecile atricapillus*), White-breasted Nuthatch (*Sitta carolinensis*) and Eastern Bluebird (*Sialia sialis*). The interspecific brood parasite, Brown-headed Cowbird (*Molothrus ater*) represents a reproductive threat to many native birds, especially open-cup grassland and shrub-nesting species such as Yellow Warblers (*Setophaga petechia*) and Dickcissels. Other common Platte Valley breeding species that the cowbird is known to parasitize include the Bell's and Warbling vireos (*Vireo bellii, V. gilvis*), Blue-gray Gnatcatcher (*Polioptila caerulea*), Yellow-breasted Chat (*Icteria virescens*), Common Yellowthroat, Grasshopper, Song (*Melospiza melodia*), and Vesper sparrows (*Pooecetes gramineus*), Eastern and Spotted towhees (*Pipilio erythrophthalmus, P. maculatus*), Northern Cardinal (*Cardinalis cardinalis*), Rose-breasted and Blue grosbeaks (*Pheucticus ludovicianus, Passerina caerulea*), Lazuli and Indigo buntings (*Passerina amoena, P. cyanea*), Red-winged and Brewer's (*Euphagus cyanocephalus*) blackbirds, Western Meadowlark and American Goldfinch (*Spinus tristis*) (Johnsgard, 1997).

Among the bird species found in the central Platte River region are several of special conservation concern as indicated by the Nebraska Natural Legacy Project (Schneider et al. 2011), the federally listed threatened Piping Plover and endangered Interior Least Tern are of particular interest. State-listed high-priority (Tier 1) At-Risk species that breed regularly in the central Platte Valley include Bell's Vireo, Burrowing Owl, Ferruginous Hawk, Greater Prairie-Chicken, Henslow's

Sparrow, and Long-billed Curlew. Of these, the Bell's Vireo and Greater Prairie-Chicken probably represent major population centers at a national level. State-listed lower-priority (Tier 2) At-Risk species that breed fairly regularly in the central Platte Valley include the White-faced Ibis, American Woodcock (*Scolopax minor*), Wilson's Snipe (*Gallinago delicata*), Bald Eagle (*Haliaeetus leucocephalus*), Swainson's Hawk, Barn Owl (*Tyto alba*), Black-billed Magpie (*Pica hudsonia*) and Swamp Sparrow (*Melospiza georgiana*). Other At-Risk species, such as the Golden Eagle, Sandhill Crane and Black-necked Stilt, are occasional or rare breeders.

As with many Great Plains rivers, the breeding bird diversity associated with the central Platte River is highest in the narrow belts of trees, shrubs and undergrowth vegetation that lie along the shorelines, collectively called riparian or riverine forests. Insects are typically abundant in these areas, as are aquatic invertebrates and vertebrates living on or in the water. Progressively farther from the river these gallery forests grade off into a variety of shrubs and small trees, such as ash (*Fraxinus* spp.), willow (*Salix* spp.), dogwood (*Cornus* spp.), and invasive species such as Russian olive (*Elaeagnis angustifolia*) (Colt, 1995; Davis, 2001; Scharf, 2007). These riverine gallery forests not only provide convenient passageways upstream and downstream for mobile animals, they serve as important breeding areas for forest-adapted birds in otherwise non-forested environments.

In the riparian forests of the 120-kilometer (75-mile) stretch of the Platte Valley between Grand Island, Hall County, and Overton, Dawson County, Davis (2005) found 56 breeding species. Neotropic migrants comprised 58 percent of the total assemblage, and 62 percent of those were classified as short-distance migrants. Forest-edge species comprised 53 percent of the total, open-forest species 25 percent and closed-forest species 16 percent. The eight most common breeding birds were House Wren, Baltimore Oriole, American Goldfinch, Blue Jay (*Cyanocitta cristata*), Common Yellowthroat, Eastern Towhee, European Starling and Northern Cardinal. The most frequently detected species were the House Wren, Baltimore Oriole, Blue Jay, Common Yellowthroat, American Goldfinch, Song Sparrow and Eastern Towhee. The breeding assemblage consists mostly of ecological generalists, having wide geographic distributions in the hardwood forests of North America. Many of the species are also fairly recent arrivals to the central Platte Valley, such as the Red-bellied Woodpecker (*Melanerpes carolinus*), White-breasted Nuthatch, Wood Duck (*Aix sponsa*) and Northern Cardinal, all of which expanded their range westward as the riparian forest matured within the past century. Only one of the species (Black-billed Magpie) was of distinctly western geographic affinities. However, during the latter 1900's the House Finch (*Carpodacus mexicanus*) moved in both directions across the central Platte Valley, and by 1990 had occupied the entire valley.

In the riparian forests between Grand Island, Hall County, and Elm Creek, Dawson County, Colt (1995) detected 50 species of breeding birds. Ecologically, 50 percent were considered to be forest-edge species, 26 percent as open-forest species, 20 percent as closed-forest species, and two percent as forest-interior species. Like Davis' findings, only one of the 50 species (Black-billed Magpie) was of distinctly western geographic affinities. The twelve most characteristic breeding species were: House Wren, American Robin (*Turdus migratorius*), "Northern Oriole" (including Baltimore and Bullock's), Song Sparrow, Blue Jay, Brown-headed Cowbird, Warbling Vireo, Gray Catbird (*Dumetella carolinensis*), Black-capped Chickadee, American Goldfinch, Northern Flicker and Yellow Warbler.

As a result of channel-cutting and sand deposition and seasonal changes in water levels and variable water flow velocity, sand bars and sandy islands regularly appear and disappear along the banks of the Platte. Over time a few plants, whose seeds were deposited on these bars and islands, germinate and gain footholds. Unless they are later swept away by ice flows in early spring or high water flows, these hardy plants, likely survive. These pioneers eventually become replaced by perennial species, whose deeper root systems are more likely to withstand the erosive effects of moving water. Eventually shrubs, and still later trees, gain ascendancy over the herbaceous vegetation, and the sandbar or island progressively comes to resemble the surrounding riparian forest (Currier, 1981; Sidle et al., 1989; Johnson, 1994).

Scharf (2007) captured birds in Dawson County from 2001–2004. Of the 90 species captured in the mainland riparian woodlands, slightly over half were also captured at in-channel woodlands and 65 were classified as locally breeding species. Scharf classified 59 as Neotropic migrants, 35 species as local or regional breeders and 27 as stopover migrants, showing the importance of this habitat type to both breeding and migratory birds. Five of the 90 species captured (Black-billed Magpie, Spotted Towhee, Black-headed Grosbeak (*Pheucticus melanocephalus*), Lazuli Bunting (*Passerina amoena*) and Bullock's Oriole are of distinctly western geographic affinities, and in all but the magpie their eastern counterpart species were also present among the captured birds.

These analyses of breeding birds in the central Platte River Valley show a high degree of similarity. Such studies have value in providing baseline data for future comparisons, as the Platte River is highly vulnerable to year-to-year variations in weather, especially rainfall. It is also vulnerable to economic pressures to extract or divert its increasingly valuable water for agriculture or other uses.

The Region's Significance as a Migratory Staging Area

There are few places in North America that, for a few weeks in March, supports as much avian biomass than does the central Platte Valley. Assume that each of the 500,000 Sandhill Cranes in the region weigh an average of six pounds, and Snow Geese (*Chen caerulescens*) are of similar weight. Between 2001 and 2003 there were estimations of from 1.2 million (2002) to 7.3 million (2001) Snow Geese and Ross's Geese (*Chen rossii*) in the Rainwater Basin and central Platte Valley, averaging 3.2 million for the three years (Vrtiska and Sullivan, 2009). Add to this a million or more Canada (*Branta canadensis*), Cackling (*Branta hutchinsii*) and Greater White-fronted geese (*Anser albifrons*) averaging perhaps five pounds, several million Mallards (*Anas platyrhynchos*) and Northern Pintails (*Anas acuta*) averaging about three pounds, and unknown numbers of other ducks of varied weights, and the total avian biomass must easily exceed 15 million pounds.

These numbers are far too large to comprehend, but the amount of food required to maintain the flocks is even more mind-boggling. Suffice it to say that, without the great quantities of unharvested grain left over in the nearby fields, such ornithological richness would be impossible. It was the great agriculture revolution following World War II, with new technology providing machinery, fertilizers, pesticides and new irrigation techniques that altered the historic Platte Valley of small farms and precipitation-dependent crops into an agricultural power-house. No wonder the cranes and geese found that the Platte Valley was the place to go to put on a maximum amount of fat in a minimum of time, prior to migrating to high-latitude breeding grounds where food is relatively scarce.

Nebraska provides the most important spring staging grounds for both Sandhill and Whooping Cranes, and the central Platte Valley contains the most important of all the state's wetlands for both species. For the Sandhill Cranes, it is the combination of the Platte River as an ideal roosting habitat, and the nearby cornfields as a convenient source of easily gathered food. For the Whooping Cranes, the Platte is less attractive as a roosting habitat than it was historically, since so much of it has been overgrown with shoreline and island vegetation. The foraging niche of Whooping Cranes is more water-based than the Sandhill's, with much of the Whooper's traditional foods obtained from shallow ponds, marshes or estuaries. Nevertheless, these cranes still regularly stop along the Platte, and sometimes feed in cornfield along with Sandhill Cranes, but they are more often found in rather remote wetlands and riparian lowlands both south and north of the Platte Valley.

Early spring along the Platte River is sheer magic. For more than a month, Sandhill Cranes pour into central Nebraska, attracted by the Platte River's abundant roosting sites and nearby meadows and fields for feeding. Looking like legions of gray ghosts from a distance, the birds often cover wet meadows and cornfields from one end to the other. Flocks of a thousand or more fly low over the river as sunset approaches, their voices rising and falling as they approach, pass overhead and disappear. With nearly a half-million cranes stopping each spring, the placid Platte is transformed into the greatest gathering of cranes in the world. Not only does the central Platte Valley shelter Sandhill Cranes in the spring, it also is a magnet to the nearly ten million ducks and geese that sprinkle the sky from horizon to horizon. Add to this the occasional appearance of a Whooping Crane or a Eurasian Common Crane (*Grus grus*), and you have the ingredients of a birding spectacle that draws thousands and thousands of people to the central Platte Valley each spring. This coming together of the river, the birds and bird-watchers is a true congruence—a word derived from the Latin *grus*, which originally meant an assemblage of cranes.

What is it that brings so many cranes to the central Platte Valley each spring? The Great Plains offer an unobstructed flyway between their Gulf Coast winter habitat and arctic breeding grounds. At a convenient mid-point in this long journey, the Platte courses east from the mountains of Colorado and Wyoming, moving slowly through farms and meadows to its eventual confluence with the Missouri River. The wide, shallow channels of the Platte provide roosting sites on bare islands and sandbars that are far away from coyotes and other land-based predators. The nocturnal protection offered by the river and the presence of spring food in the form of waste corn and invertebrates are key attractions for cranes.

Cranes as a taxonomic group have existed for more than 50 million years, and the 16 species of these large wading birds are scattered across most of the continents, of which two occur in North America. Evidence of their presence in this region some nine million years ago exists in the form of the single fossilized wing bone of an ancestral crane, found along a creek in what is now western Nebraska. Nobody knows whether these birds nested or roosted along a river resembling the modern Platte River. Spring in the central Platte Valley begins when Sandhill Cranes first appear in the sky, after finishing 400-mile flights from their winter homes in Texas and New Mexico. They appear when a unique combination of warming breezes out of the south, melting ice, and blue skies are transformed into hundreds of clarion crane calls from high overhead. There is a majesty and ballet-like beauty in the synchronized movements of hundreds of birds all simultaneously setting their wings, lowering their legs and parachuting downward, as if they knew they had finally returned home. There is also a slowly growing crescendo of their calls, as flock after flock drops into the river. Later, as the sun sinks slowly below the horizon, the silvery gray birds are slowly transformed into black silhouettes as they rest from their long passage.

Each evening near sunset, flocks of a thousand or more cranes fly low over the river, their voices rising and falling as they approach, pass overhead and disappear again in the distance. As the sun sinks toward the horizon the birds become increasingly nervous, hoping to find a safe landing place before it is wholly dark. A lone, brave crane then touches down, followed moments later by another, then dozens and finally hundreds. After some initial jostling for position and rejoining of any pairs or family members separated in the confusion of landing, darkness settles on the crane roost. But even during the darkest hours it is never completely quiet, as nighttime conversations suggest that a few birds are always awake and on the alert for danger.

As darkness gives way to dawn, the volume of crane conversation gradually builds. Typically, small groups of Snow, Canada, Cackling and Greater White-fronted geese begin to leave their river roosts before the cranes, but with the first appearance of the sun the cranes start to stir. Then, sometimes as a single amorphous mass, an entire roost of several thousand birds may take flight, their combined notes shattering the dawn like the thunder of a sudden electrical storm. The flocks head out from the river toward their favorite foraging sites. Within half an hour after sunrise, the river is usually devoid of cranes, leaving it free for the Common Mergansers (*Mergus merganser*), Mallards and Northern Pintails to feed and conduct their courtship rituals. The cranes spend most of the day in the cornfields and meadows harvesting whatever grain and invertebrates they can find, building up fat stores essential for their remaining spring migration and arctic breeding.

After five or six weeks of regular cycles of daytime foraging and nightly roosting, the pattern breaks up. During a sun-warmed day in mid-April, when thermals are forming and a gentle breeze comes out of the south, the birds ascend in great slow-motion whirlwinds, their wings lifted by the invisible thermals until the cranes are almost out of sight. Soon after leaving the Platte Valley, the great flocks begin to split up, some heading for Hudson Bay shorelines and islands, and others for the high-arctic tundras of far-northern Canada. Still others head to the Yukon-Kuskokwim delta of Alaska and others to Siberian tundras some 3,000–4,000 miles away. They may not arrive at these nesting grounds until the end of May or early June, just as these areas are becoming snow-free.

Some of the larger Sandhill Cranes that pass through the eastern end of the central Platte Valley (those of the "greater" race) nest in the Great Lakes region, west to Ontario, Minnesota and eastern Iowa. Greater Sandhill Cranes once even nested as far south as the marshlands of the Nebraska Sandhills where long breeding seasons and abundant food must have provided an ideal situation. But hunting pressure and disturbances caused by ranching and farming in the Nebraska Sandhills eventually pushed nesting cranes out of Nebraska by the late nineteenth century. There is hope that the cranes will eventually reclaim the Sandhills marshes for nesting as some pairs have been sighted in the Sandhills.

An unknown percentage of the Sandhill Cranes passing through Nebraska are intermediate in size between the lesser and greater races, and have been named the Canadian race. They mostly nest in the boreal forests of Canada, geographically between the tundra-nesting lessers and the open woodland and prairie-nesting greaters. These cranes have been studied very little, and biologists are often at odds as to whether or not they represent a distinct and recognizable subspecies.

The Whooping Crane is the gold standard of American birds: it is the tallest, one of the rarest, and certainly one of the most beautiful of all North American birds. Somewhat like Sandhill Cranes, Whooping Cranes migrate from their subtropical winter habitat on the Gulf Coast to subarctic breeding grounds in the north. But unlike Sandhill Cranes, Whooping Cranes move in small flocks, with family-sized units the norm. They also are more water-dependent than Sandhills, preferring to forage in wetlands rather than uplands, and seeking out roosting sites with wider stretches of river than those accepted by Sandhill Cranes.

Historically, perhaps the entire Whooping Crane population, possibly as many as 1,000, moved up the Central Flyway each spring, many of them probably stopping at the Platte River. A survey of Whooping Crane sightings recorded during the early to middle 1900's indicated that the "Big Bend" region of the central Platte Valley was their most important stopover point. But as the Platte and its associated wetlands declined and deteriorated from human activities, the Whooping Cranes' total population dropped to about 20 birds, and they stopped coming regularly to the Platte.

In recent decades, as Whooping Crane numbers have slowly recovered, they have increasingly used the Rainwater Basin wetlands south of the Platte Valley, They has also used some of the treeless rivers and wet meadows in the Nebraska Sandhills, and wetlands as far north as the Niobrara River. These changes in Whooping Crane distribution in Nebraska almost certainly are the result of the "dewatering" of the Platte River by agricultural and residential water users during the twentieth century. The river has lost much of its historic flows to irrigation, eliminating many of its myriad channels, reducing its width, drying up the grassy meadows along its shorelines and allowing woody vegetation to grow on its margins and sandbars.

Sandhill Cranes have adjusted to these river changes fairly well, but the rarer birds breeding on the river's sandy bars and islands—such as Piping Plovers and Interior Least Terns—have not fully recovered. Efforts by Audubon's Rowe Sanctuary, The Crane Trust, and the Platte

River Recovery Implementation Program, established largely to preserve these imperiled species and their habitats, have slowed and locally reversed these long-term changes in the Platte's ecology, but much of the damage will be difficult to repair.

Despite the challenges facing the Platte, the annual crane migration spectacle remains enchanting. From the middle of February through the end of March, the number of Sandhill Cranes slowly increases, reaching a peak near the end of March. Then, during the first two weeks of April, when south breezes and sunshine generate thermal updrafts over the plains, flock after flock lifts off and heads resolutely northward. After they have gone, the Platte reverts to being an ordinary Great Plains river, waiting silently for another congruence of the cranes.

Regionally Threatened and Endangered Birds

The Whooping Crane

Although many Nebraskans have had the indescribable pleasure of watching thousands of Sandhill Cranes overhead or roosting on Platte River sandbars, relatively few have seen Whooping Cranes. Compared with approximately 500,000 Sandhill Cranes migrating through the state each year, there were as of 2012 less than 300 Whooping Cranes in the flock that migrates from Aransas National Wildlife Refuge, on Texas's Gulf Coast, to Wood Buffalo National Park, on the border of Alberta and Canada's Northwest Territories. Some 245 Whoopers wintered at Aransas during the winter of 2011–2012, down from 283 the year before; but an uncertain number of others wintered outside the refuge, owing to a combination of unfavorable ecological conditions. Whooping Cranes migrate somewhat later in spring than the Sandhills (during April in Nebraska). During daytime foraging they usually frequent remote wetlands far from any roads, and generally move in small groups of pairs, family, or extended families that often consist of a pair and one or more generations of their offspring.

The attraction of cranes to many people has meant that we now know as much about the lives of Sandhill and Whooping cranes as almost any other North American bird. The Whooping Crane was listed as a nationally endangered species in 1972. By 1941 only 22 Whooping Cranes survived (16 in Texas and six in coastal Louisiana) in the wild. The Louisiana population was extirpated in 1949. It was not until 1954 that the species' Canadian breeding grounds were discovered, and not until 1986 that their world population reached 100 individuals. It is because of the Whooping Crane's perilous status that the Platte River has been protected from destruction, through the identification of the central Platte Valley as critical habitat for the species. Thereby, the Whooping Crane's threat of extinction has helped preserve Platte habitats for many other water-dependent species. The central Platte Valley was the first area federally designated as critical habitat for migratory Whooping Cranes between Texas and the Canadian border. Other Great Plains sites that have been similarly identified as critical habitats include Oklahoma's Salt Plains National Wildlife Refuge, and Cheyenne Bottoms State Wildlife Area and Quivira National Wildlife Refuge, both in Kansas. The recognition of the Platte's importance to the survival of the Whooping Crane also resulted in the establishment of a habitat mitigation fund associated with the building of the Grayrocks Dam on a major Platte tributary (the Laramie River) in eastern Wyoming, and the formation in 1978 of the Platte River Whooping Crane Habitat Maintenance Trust (renamed the Crane Trust in 2012).

In Nebraska, the peak spring migration period is early April, and the peak fall migration is mid-October. Starting in 1979, individual Whooping Cranes have been arriving in Nebraska much earlier was previously was the case, possibly because they become socially attached to Sandhill Crane flocks that now typically arrive in the Platte Valley by mid-February.

In order to better document the role of individual cranes in migration, their strong family affinities, great longevity and the species population structure, a program of color-banding juveniles was initiated in 1977 and continued until 1988. Twenty-four of the 132 banded birds were still alive in 2009; all survivors were at least 22 years old, and one had reached 32 years of age. Many of these birds have continued to use the very same migratory stopover locations at did their great-grandparents, showing the power of place-memory in crane migration, and the importance of migratory traditions in long-lived and long-distance migrants.

Accidents are probably the major cause of adult mortality in Whooping Cranes, especially collisions with overhead utility lines by migrating birds. By comparison, sport hunting is the cause of most mortality in Sandhill Cranes, resulting in the deaths of more than five percent of the lesser Sandhill Crane population annually. Hunting for Sandhill Cranes in Nebraska has never been allowed, because of the special importance of the Platte River to Whooping Cranes and the dangers of hunting-related mortality. Whooping cranes have been shot by hunters in recent years in both South Dakota (2011) and Kansas (2004), as well as in several other states, in spite of continued hunter-education efforts.

For the biologists who have worked so hard to restore Whooping Cranes such disasters as losing birds to hunting or accidents are heart-breaking, and likewise even small miracles such as the gaining of a few chicks per year are rewarding. By 2012 the mid-continent population of Whooping Cranes had reached nearly 300 birds, in spite of an extended drought on the southern Great Plains. With drought, floods and human activities impacting wetlands and rivers throughout the Great Plains, the thought of Whooping Cranes still flying high overhead is comforting, and one that we must all act in such a way as to keep natural treasures such as cranes a continuing reality.

The Eskimo Curlew

Shortly after daylight on September 16, 1932, four shorebirds flew in from the direction of the sea and dropped into the vegetation near the lighthouse on Montauk Point, on Long Island, New York. Stalking the birds, ornithologist Robert Cushman Murphy could scarcely believe his eyes, they were almost certainly Eskimo Curlews (*Numenius borealis*), a species believed by some to be extinct. They had probably flown in from the coast of Labrador, where a curlew had been shot two weeks previously near the Strait of Belle Isle, so far as is known, that was the last Eskimo Curlew ever shot on the North American continent, and Murphy's observation was to be the last along the Atlantic coast for nearly thirty years.

Those four birds were representatives of a species once so abundant that flocks were reported to be a mile long and nearly as wide. When the pre-migratory birds arrived in Labrador from central Canada in August, they gorged themselves on invertebrates and crowberries. The birds remained in Newfoundland and Nova Scotia until early September, when they would leave for a nonstop flight to the Lesser Antilles. After a brief stop in the Lesser Antilles, the birds continued south over eastern Brazil and on to Argentina. They arrived at their winter quarters by mid-September, concentrating in the grassy pampas south of Buenos Aires. Fall storms on the Atlantic coast affected this schedule and itinerary, often forcing the birds to hug the North American shoreline. Large flocks would build up along the Atlantic coast, particularly in Massachusetts, on Long Island, and down through the Middle Atlantic and southern states. Under adverse weather conditions, some flocks sought refuge on Bermuda and westerly winds sometimes drove them far out over the Atlantic.

The curlew's spring migration route was quite different from the fall pattern. By late February or early March the birds left their wintering areas and undertook a nonstop flight to the US Gulf Coast. They concentrated in the coastal prairies of Texas and gradually moved northward through the Great Plains. Birds began arriving in Oklahoma in late March, and by early April they were abundant in Kansas, Missouri, Iowa, Nebraska, and Oklahoma. They foraged and migrated in vast flocks; in flight these might reach a halfmile in length and a hundred yards or more in width. When feeding in fields they frequently covered areas of 40–50 acres. These great spring flocks reminded observers of Passenger Pigeons (*Ectopistes migratorius*), and the birds were often called "prairie pigeons." The largest flocks assembled on newly plowed fields or burned prairie, searching for grasshopper eggs, young grasshoppers, and other insects. Market hunters ravaged these flocks. When they were shot for sport, the bodies were simply dumped on the prairie forming heaps as large as several tons of coal. The birds circled in masses so dense that one "could scarcely throw a brick or other missile into [them] without striking a bird," according to Myron Swenk (1915).

Nebraska seems to have been one of the curlews' major spring staging areas. There is no evidence of the birds stopping in North Dakota or in the Prairie Provinces of Canada, they apparently left South Dakota in mid-May, made an almost nonstop flight to the Northwest Territories, and reached the vicinity of Great Slave Lake in the late May. There is no firm evidence that they nested in Alaska, but small curlews were seen in the late 1800's at Saint Michael and Cape Lisbourne in May, and they were seen near Point Barrow between May 20 and July 6, 1882.

By the middle of August the curlews began their fall migration. The majority of birds, presumably the adults, flew eastward across the northern part of Hudson Bay, making for the Labrador coast. The birds that reached Labrador by the middle of August were probably unsuccessful breeders. There was apparently a second migration route down the west coast of Hudson Bay that reached the Atlantic coast via Ontario and the Great Lakes. A few birds also moved southward through the Great Plains, essentially retracing their spring migration route, these may have been young birds. The few fall records for Nebraska suggest that Eskimo Curlews passed through state in October.

In Nebraska, the last large flock (70–75 birds) was observed in Merrick County in April 1900, while other, smaller flocks were seen in York County in 1904 or 1905 and in Madison County in 1909 or 1910. Among the last birds shot were seven (out of a flock of eight) killed on April 19, 1915 in Merrick County. A single bird was killed on April 17, 1915, near Norfolk, Madison County, this may have been the last curlew killed in Nebraska. A group of eight birds was observed near Hastings on April 8, 1926. A single curlew was reportedly seen on 16 April 1987 at Crane Meadows, Hall County, Nebraska, however the report is not considered reliable (Faanes 1990; Grenon 1991). In 1924 a small flock was seen near Buenos Aires, Argentina, and one was collected; the next year a single bird was shot in the same location. In 1929 one was killed in Maine. A collector shot one bird in Labrador in 1926, four in 1927, and one in 1932. In 1945 two were reportedly seen in late April on Galveston Island, Texas, and another was reported in mid-July 1956 on Folly Island off the South Carolina coast. In June 1946 there was another sighting on the South Carolina coast, and in April 1950 an observer reported seeing a single bird near Rockport, Texas. In September 1964, a specimen was killed on Barbados.

On March 22, 1959 two observers saw a small, strange-looking curlew feeding among a group of Long-billed Curlews. The bird was seen until April 12, and in 1960 and 1961, perhaps the same bird was observed and successfully photographed, and in 1962 at least three and possibly four curlews were present on Galveston Island. These sightings added to the hopes that a breeding population might still exist. Countering this optimism was the possibility that the birds may have been Asian Little Curlews (*Numenius minutus*).

There were few sightings in the period between 1965 and the present. In August 1970, a single bird was reported at Plymouth Beach, Massachusetts. In August 1972, two birds were observed on Martha's Vineyard, Massachusetts. In August 1976, a pair was seen flying over an area of coastal tundra along the west coast of James Bay. Between 1932 and 1976, there were at least six sightings the Eskimo Curlew on the coast of Texas, seven from the Atlantic coast, and one from James Bay. Thus, there was some evidence that this species, believed to have been extinct by 1920, probably still existed.

Most ornithologists who have considered the Eskimo Curlew's story have come to the conclusion that uncontrolled hunting was the most important factor in the species' demise. Banks (1977) summarized the possible reasons why the Eskimo Curlew could nearly go extinct over a period of only a few decades. He suggested that sport and market hunting pressures late in the 19th century were unlikely to have been the sole mortality factor; other possible factors included habitat changes, persistent breeding failures and global climatic changes influencing storm frequencies and intensities. Banks (1977) concluded that a major cause of the species' decline was the intensification of storms associated with warming global conditions late in the 19th century. Storms occurring during the overseas migration from the Atlantic Coast to South America could have devastated flocks and a single fall hurricane could have destroyed many of the last surviving birds. Another cause of the decline was increased commercial hunting along the East Coast, beginning in the 1880's. Interestingly, American Golden-Plovers and Hudsonian Godwits (*Limosa haemastica*), which have similar migration patterns as the Eskimo Curlew and were similarly adversely affected by market hunters, also suffered major population declines during this same time period.

The Interior Least Tern

Interior Least Terns were one of the first species of bird to greet explorers as they ventured into the heart of the Great Plains. On 5 August 1804, Meriwether Lewis and William Clark on their 1803–1805 "Voyage of Discovery" across the continent, saw terns along the Missouri River in present-day Washington County. Clark wrote that "this bird is very nosey when flying which it dose extremely swift the motion of the wing is much like that of Kildee it has two notes like the squaking of a small pig only on reather a higher kee, and the other kit'-tee'-kit'-tee'—as near as letters can express the sound" and "a native of this country and probably a constant resident" (Moulton 2005). John James

Audubon, Ferdinand Hayden, traveling with the Maj. Stephen Long Expedition of 1819–1820, and Paul Wilhelm, Duke of Wurttemberg, traveling in 1823, all reported finding terns in large numbers nesting on sandbars in the Platte and Missouri rivers in the region that would become Nebraska. Hayden found terns and plovers nesting on sandbars near the confluence of the Loup and Platte rivers, an area then known as the Loup Fork, and now considered the eastern edge of the central Platte River Valley (Ducey, 2000).

Interior Least Terns are among the smallest species of the tern family, and about the size of an American Robin (*Turdus migratorius*). Adults of both sexes have identical plumages, with a black-capped crown and white forehead patch, gray back and wings, white underparts and orange-yellow beak and legs. They nest in small colonies; laying 2–3 well-camouflaged eggs in small, cuplike depressions in bare sand. It is not unusual to find a rock or stick near the nest, perhaps serving as a windbreak for the incubating bird, or possibly as a nest landmark. Incubation lasts about 21 days, and, after being fed fish and parented for another 21 days, the chick fledges and gradually becomes independent of its parents. Adult terns try to protect their nests and chicks by swooping down and "dive-bombing" any intruder.

The Interior Least Terns found in the central Platte River Valley are one of the three subspecies found in the New World (Interior, *S. a. athalassos*; eastern or coastal, *S. a .antillarum*; California, *S. a. browni*; AOU, 1998). Historically the Interior race nested on sandbars in the Mississippi, Missouri, Ohio, Arkansas, Red, and Rio Grande River systems and the rivers of central Texas. Their range extended from Texas to Montana and Colorado and New Mexico to Indiana. Terns continue to breed in these river systems, but their distribution is restricted to less-altered reaches. The exact overwintering locations of the central Platte River population of Interior Least Terns are unknown, but Least Terns are found during winter along the Central American coast and the northern coast of South America from Venezuela to northeastern Brazil (Sidle and Harrison 1990; Thompson et al., 1997). The Interior Least Tern was placed on the federal Endangered Species List (endangered) on 27 June 1985 after an 11-year evaluation process, and the recovery plan was released in 1990 (Sidle and Harrison 1990).

The Piping Plover

Lewis and Clark also observed Piping Plovers on their 1804–1806 travels through the Great Plains, referring to them in their journals as 'small kildee' (Moulton 2005). Ferdinand Hayden considered plovers to be abundant across the region during the 1850's, and found them nesting on sandbars in the Platte River (Ducey, 2000). Native Americans in the Great Plains included them in their stories and legends; *ūt* refers to plover in the Pawnee language (Ducey, 2000).

Piping Plovers are robin-sized birds, with buffy-gray backs and wings, white underparts, and orange legs and beaks. Like terns, male and females have identical plumages. Their most distinctive plumage feature is the single necklace-like black band across their upper breast. Plover nests resemble tern nests in that they are small, cuplike depressions in sand. Nests nearly always hold four eggs, and are lined with small, light colored pebbles. Incubation lasts about 28 days. Adult plovers feign a crippled-wing behavior to lure predators away from their nests and chicks. After being parented for 28 days the chick fledges (Haig et al., 1988; Haig, 1992).

The Piping Plovers found in the central Platte River Valley are part of the Great Plains population that ranges from Alberta to Manitoba and south to Nebraska; there is a very small population nesting around the Great Lakes, and a population nesting along the North American Atlantic Coast from Newfoundland to North Carolina. The Atlantic Coast population has been designated as *C. m. melodus* and the inland population *C. m. circumcinctus*. However, the validity of the latter subspecies status is questionable. The wintering range of Piping Plovers has become better known through a range-wide color-banding program. The majority of the plovers in Nebraska's Platte River population migrate directly south to the beaches and sandflats along the Gulf Coast, mostly in the area between Aransas and Corpus Christi, Texas. Some Nebraska-hatched plovers have also been reported from beaches near New Orleans, Louisiana, Mobile, Alabama, and from Pensacola and Ft. Myers, Florida. Birds from the Atlantic Coast population winter in the Bahamas and along the east coast of Florida (Haig et al., 1988; Haig, 1992).

The Piping Plover was placed on the federal Endangered Species List (threatened) on 10 January 1986, after a four-year evaluation process. A recovery plan was completed in 1988 (Haig et al., 1988). Critical breeding habitat for the Great Plains population was designated in Montana, Nebraska, South Dakota, and Minnesota in 2002. However, as the result of a legal challenge by the Nebraska Habitat Conservation Coalition in 2005 the U.S. District Court vacated the segment of critical habitat in Nebraska.

The populations of Interior Least Terns and Piping Plovers nesting in the central Platte River Valley and across the Great Plains have declined for reasons directly related to human activities. Nineteenth-century market hunting and hunting for the millinery trade had much less impact on the populations in the Great Plains than did those activities on the Atlantic and Gulf coasts. Since then, the broad-scale alteration of river flows caused by water diversions for agricultural, municipal, or commercial purposes have resulted in loss of their sandbar nesting habitat, and similar broad-scale loss of beach and sandflat wintering habitat due to commercial and recreational development along their wintering grounds.

Continuing threats to both species' nesting sandbar habitats during the breeding season include (1) the presence of invasive plants, such as common reed (*Phragmites australis*), purple loosestrife (*Lythrum salicaria*), and salt-cedar or tamarisk (*Tamarix* spp.); (2) the construction of dams and reservoirs controlling river flows and sandbar creation; (3) river channelization and bank stabilization activities, changing the physical nature of the river; (4) hydropower generation ("hydro-peaking"), causing sandbars to be overtopped and nests washed away; and (5) water diversion for agricultural and municipal use, causing the "dewatering" of the river. Current threats to wintering habitat include (1) rising sea levels due to global climate change, eliminating sandy beaches and sandflats, (2) wind-power development along coastlines that interfere with migratory flight-lines, and (3) commercial, residential and recreational development along coastlines.

Interior Least Terns and Piping Plovers often nest on artificial habitats such as sand and gravel mines, dredge islands, lake shore housing developments, man-made sandbars, reservoir shorelines, ash disposal piles at coal fired power plants, and, in rare instances, on tar-gravel rooftops (Sidle and Harrison, 1990). The frequency of nesting in these artificial habitats varies, depending upon the amount of midstream sandbar-nesting habitat available to the birds. In recent years, the hab-

itat restoration activities of the Platte River Recovery Implementation Program, the Crane Trust, and Rowe Audubon Sanctuary have increased the amount of sandbar habitat available in the central Platte
River Valley. The legal protection of Interior Least Terns, Piping Plovers, and Whooping Cranes is mandated under the multiple auspices of
the 1917 International Migratory Bird Treaty Act, the 1973 U.S. Endangered Species Act, and the 1975 Nebraska Nongame and Endangered
Species Conservation Act.

Part II.

Annotated List of Regional Birds

The following species accounts are based on Nebraska Ornithologists' Union records (NOU), seasonal occurrence reports in the Union's quarterly journal *Nebraska Bird Review*, other published reports, personal observations, and miscellaneous records. Species are arranged taxonomically according to the American Ornithologists' Union's (AOU) *Check-list of North American Birds* (1998), 7th edition, with supplements 42–52 through 2012 (available at http://www.aou.org/check-list). Currently accepted English and Latin names follow AOU naming conventions. Abundance categories are based on the following criteria: *abundant*, found in large numbers, often in flocks; *very common*, certain to be seen in numbers in suitable habitat; *common*, certain to be seen in suitable habitat; *uncommon*, likely present but not certain to be seen; *occasional*, not seen regularly, but expected during the appropriate season; *rare*, seen only a few times during a season; *very rare*, not normally seen and considered to be outside the typical range; and *hypothetical*, questionable record or likely the result of escape from captivity. Seasonal occurrence is based on the following calendar: *winter*: December-January-February, *spring*: March-April-May, *summer*: June-July-August, and *fall*: September-October-November.

In 2005, the Nebraska Natural Legacy Project developed a list of species that are at risk of extirpation or extinction from the state. Tier 1 species are those identified as endangered or threatened under the Nebraska Non-game and Endangered Species Conservation Act (NESCA), and/or the federal Endangered Species Act, are globally ranked as critically imperiled, imperiled, or vulnerable by NatureServe and the Natural Heritage Network, or whose abundance and/or distribution has declined across their range. Tier 2 species are those identified as state critically imperiled, imperiled, or vulnerable. The Legacy list is reviewed and revised regularly; the species' status categories listed here (Schneider et al. 2011) reflect the 2010 revision. A total of 13 of Nebraska's 21 Tier 1, and 58 of Nebraska's 84 Tier 2 At-Risk species are found in the region covered by this study.

Species status categories are also provided for the following conservation entities: (1) the threatened and endangered categories of the federal Endangered Species Act (ESA), (2) the American Bird Conservancy's Red-List and Yellow-List species (ABC; http://www.abcbirds.org), (3) the United States Shorebird Conservation Plan conservation categories (USSBCP; Brown et al., 2001), and (5) the Canadian Shorebird Plan conservation categories (CSBP; Donaldson et al., 2000).

The following list includes 372 species recorded through 2011 in the central Platte River Valley and nearby counties as defined earlier. Available county occurrence records (published and unpublished) are listed at the end of each species account, using the following abbreviations: Adams (A), Buffalo (B), Clay (C), Dawson (D), Frontier (F), Gosper (G), Hall (HL), Hamilton (HM), Kearney (K), Lincoln (L), Merrick (M), and Phelps (P). Other regional abbreviations identify federally-owned Waterfowl Production Areas (WPAs) and state-owned Wildlife Management Areas (WMAs). Among species with two molts and plumages annually the "alternate" plumage refers to the breeding plumage, and the "basic" plumage to the non-breeding or winter plumage. Species described as "visitors" are not known to breed in the region; "residents" are known or assumed to breed there.

Records listed for the Pink-footed Goose (*Anser brachyrhynchus*), Emperor Goose (*Chen canagica*), Swan Goose (*Anser cygnoides*), Barnacle Goose (*Branta leucopsis*), Mute Swan (*Cygnus olor*), Ruddy Shelduck (*Tadorna ferriginea*), Chukar (*Alectoris chukar*), Coturnix Quail (*Coturnix coturnix*), White-tailed Tropicbird (*Phaethon lepturus*), Black Vulture (*Coragyps atratus*), Budgerigar (*Melopsittacus undulates*), Monk Parakeet (*Myiopsitta monachus*), Acadian Flycatcher (*Empidonax virescens*), and Gray-crowned Rosy-Finch (*Leucosticte tephrocotis*) are considered as representing intentional but failed releases, unintended escapes from captivity, or poorly documented observations; these accounts are shown in brackets ([ ]) and are not included in the species total.

Further information on the birds and natural history of the region can be found in Bruner et al. (1904), Tout (1902, 1947), Lingle (1994), Mollhoff (2001), Sharpe et al. (2001), Jorgensen (2004, 2012), Freeman

(2010), and Johnsgard (2005, 2008, 2011, 2012). Additional seasonal and distributional information is available in the reports compiled by the Nebraska Ornithologists' Union (NOU) and published in the *Nebraska Bird Review*, as well as various internet sources of bird records, such as eBird (http://ebird.org/content/ebird/) and NEBirds (http://groups.yahoo.com/invit/nebirds).

1. Non-passerine Orders

Order Anseriformes

Family Anatidae: Geese, Swans, and Ducks

Black-bellied Whistling-Duck, *Dendrocyna autumnalis*
Rare. The first state record is represented by a bird shot at Hansen WPA October 28, 1969 (NBR 58:49–52). A pair was recorded at North Harvard Basin, Clay County August 2, 1999, and, probably the same pair at Hultine WPA, Clay County August 21, 1999 (Jorgensen, 2012); also reported from Harvard WPA, Clay County June 6–7, 2010 (NBR 78:88).
County Records: [C, HA, P]

Taiga Bean-Goose, *Anser fabalis*
Very rare. Reported observed at Funk WPA, Phelps County April 4, 1998 (Sharpe et al., 2001).
County Records: [P]

[**Pink-footed Goose**, *Anser brachyrhynchus*]
Hypothetical. One bird was photographed January 30, 2006, at Harvard WPA, Clay County (Jorgensen, 2012). This record was not accepted by the NOU records committee, owing to the bird's uncertain provenance.
County Records: [C]

Snow Goose, adults of white and blue plumage morphs.
This and all following illustrations by P. Johnsgard.

Greater White-fronted Goose, *Anser albifrons*
Common spring and fall migrant throughout the region, to be expected from early March to mid-April and from mid-October to
mid-November.
County Records: [A, B, C, D, F, G, HL, HM, K, L, M, P]

[**Swan Goose**, *Anser cygnoides*]
Hypothetical. Observed April 4, 2000 in Phelps County. This is a
commonly raised exotic species, and the record is considered to
represent an escaped captive (NBR 69:90).
County Records: [P]

[**Emperor Goose**, *Chen canagica*]
Hypothetical. One adult was found dead during an avian cholera
outbreak at Harvard Lagoon, Clay County March 17, 1997 (NBR
65: 77). The bird showed no signs of captivity and the NOU determined it to be a wild bird (Brogie, 1998).
County Records: [C]

Ross's Goose, adult

Snow Goose, *Chen caerulescens*
Abundant migrant throughout the region, to be expected from early March to mid-April, and from early October to early December.
County Records: [A, B, C, D, F, G, HL, HM, K, L, M, P]

Ross's Goose, *Chen rossii*
Uncommon migrant throughout the region, to be expected from early March to mid-April, and from early October to early December.
County Records: [A, B, C, D, F, G, HL, HM, K, L, M, P]

Brant, *Branta bernicla*
Rare spring migrant. Observed at Funk WPA, Phelps County, March 21, 1989, from Kearney County, February 22, 1998 and at Lake Hastings, Adams County, March 6–7, 2011 (Jorgensen, 2012). *B. b. bernicla* occurs in the region, and may travel with migrating Canada and Snow geese.
County Records: [A, B, K, P]

[**Barnacle Goose**, *Branta leucopsis*]
Hypothetical. One bird seen in a sandpit lake near Odessa, Phelps County, March 9, 1995. The NOU did not consider this to be a wild bird and did not accept the record (Gubanyi, 1996).One was seen at Massie Lagoon, Clay County, March 28 to April 4, 1998; what may

have been the same bird was seen at Harvard Lagoon, Clay County May 9, 1998 (Sharpe et al., 2001).
County Records: [C, P]

Cackling Goose, *Branta hutchinsii*
Common to abundant spring and fall migrant throughout the region, to be expected from early March to mid-April, and from early October to early December. Possibly overwinters locally, but the winter status of this species in the region is unresolved. Current evidence indicates that the central Platte River Valley is a major migratory corridor for this smallest of the "white-cheeked" geese of central North America (Johnsgard, 2012a, 2012b).
County Records: [A, B, C, D, G, HL, HM, K, L, M, P]

Canada Goose, *Branta canadensis*
Very common to abundant migrant and local permanent resident throughout the region. The species has become increasingly common as a breeding bird in Nebraska since restoration efforts of the latter 1900's. It tends to be fairly sedentary, and often overwinters on ice-free waters.
County Records: [A, B, C, D, F, G, HL, HM, K, L, M, P]

Trumpeter Swan, *Cygnus buccinator*
Rare spring and fall migrant, and a local winter resident in adjacent southwestern Sandhills. Regionally reported from November to March (Lingle, 1994), and also June (Jorgensen, 2012). A bird collected near Overton, Dawson County on March 15, 1898 is the first record for the state (Sharpe et al., 2001). These swans overwinter on nonfreezing artesian-fed creeks flowing south out of the Sandhills into the North Platte and Platte river valleys. Tier 1 At-Risk and ABC Yellow-List species.
County Records: [C, D, F, HL, L]

Tundra Swan, *Cygnus columbianus*
Rare migrant, mainly east, to be expected from March–April and November. A bird collected near Doniphan, Hall County October 27, 1917 is the first record for the state (Sharpe et al., 2001).
County Records: [A, C, HL]

[Mute Swan, *Cygnus olor*]
Hypothetical. All records of this common zoo and park species are presumably the result of released or escaped birds. Three birds

were seen near Odessa, Buffalo County on June 5–8, 1997 (Sharpe et al., 2001).
County Records: [B]

[**Ruddy Shelduck**, *Tadorna ferriginea*]
Hypothetical, one bird seen in Kearney, Buffalo County on February 24, 1997, considered an escaped bird based on its behavior (Sharpe et al., 2001).
County Records: [B]

Wood Duck, *Aix sponsa*
Very common migrant and uncommon summer resident, to be expected from late March to late October.
County Records: [A, B, C, D, F, G, HL, HM, K, L, M, P]

Wood Duck, adult pair

Gadwall, *Anas strepera*
Common migrant and local summer resident throughout the region, to be expected from late March to November.
County Records: [A, B, C, D, F, G, HL, HM, K, L, M, P]

Eurasian Wigeon, *Anas penelope*
Rare migrant. There are April and May regional records (Lingle, 1994), and Jorgensen (2012) noted Rainwater Basin records from March 5 through April 27. A bird collected at Inland, Clay County September 26, 1914 is the first record for the state (Sharpe et al., 2001). Usually occurs with American Wigeon, and several apparent Eurasian x American Wigeon hybrids have been observed in the region (NBR 62:69). Few female and fall records exist, owing to the similarities of the species.
County Records: [A, C, D, HL, HM, K, L, M, P]

American Wigeon, *Anas americana*
Very common migrant throughout the region, and a very rare summer resident, to be expected from mid-March to early May, and from late September to mid-November. There is a 1962 report of nesting in Lincoln County (NBR 30:24–25) and Jorgensen (2012) reported a brood at Kissinger Basin WMA, Clay County, June 30, 2001. Tier 2 At-Risk species.
County Records: [A, B, C, D, F, G, HL, HM, K, L, M, P]

American Black Duck, *Anas rubripes*
Rare fall and winter migrant. There are December to February regional records (Lingle, 1994). There is a specimen record from near Sutton, Clay County, September 20, 1910, and a sighting at Harvard WPA, Clay County January 9, 2006 (Jorgensen, 2012). American Black Duck x Mallard hybrids are possible in the region, and are quite common farther east.
County Records: [B, C, HL]

Mallard, *Anas platyrhynchos*
Very common to abundant migrant and summer resident throughout the region, often overwintering.
County Records: [A, B, C, D, F, G, HL, HM, K, L, M, P]

Blue-winged Teal, *Anas discors*
Very common migrant and local summer resident throughout the region, to be expected from early April to mid-October.
County Records: [A, B, C, D, F, G, HL, HM, K, L, M, P]

Gadwall (*above*) and American Wigeon (*lower*), adult males

Cinnamon Teal, *Anas cyanoptera*
Uncommon migrant and highly local summer resident, in the west, to be expected from April to October. Reported regionally from March to May (Lingle, 1994) but difficult identification makes fall records questionable. Tier 2 At-Risk species.
County Records: [A, B, C, D, F, HL, K, L, M, P]

Northern Shoveler, *Anas clypeata*
Very common migrant and highly local summer resident throughout the region, to be expected from March to November.
County Records: [A, B, C, D, F, G, HL, HM, K, L, M, P]

Northern Pintail, adult male

Northern Pintail, *Anas acuta*
Very common migrant and local summer resident throughout the region, often overwintering.
County Records: [A, B, C, D, F, G, HL, HM, K, L, M, P]

Garganey, *Anas querquedula*
Very rare spring migrant. A male in alternate plumage was seen near Kearney, Kearney County March 28, 1997 (Brogie, 1998; NBR 65:77).
County Records: [K]

Green-winged Teal, *Anas crecca*
Very common migrant and highly local summer resident throughout the region, to be expected from mid-March to early November. A nest was found at Massie WPA, Clay County, in 1985 (Harding, 1986; NBR 68:110), and a brood was seen at Harvard WPA, Clay County in July, 2007 (Jorgensen, 2012).
County Records: [A, B, C, D, F, G, HL, HM, K, L, M, P]

Canvasback, *Aythya valisineria*
Common spring and fall migrant, rarely overwintering or summering in the region. Reported regionally from October to December, and from late March to May with a few summer occurrences, and from August to December, including Rainwater Basin records (Lin-

gle, 1994; Jorgensen, 2012). Tier 2 At-Risk species.
County Records: [A, B, C, D, F, G, HL, HM, K, L, M, P]

Redhead, *Aythya americana*
Very common spring and fall migrant throughout the region, rarely overwintering or summering. To be expected from mid-March to mid-November, but reported regionally by Lingle (1994) for all months from November to June except for January. A single January record is known (Jorgensen, 2012). Found nesting at Harvard WPA, Clay County, June 2, 2007 (NBR 75:73). There is an historic (1916) nesting record for Clay County (Ducey, 1988).
County Records: [A, B, C, D, F, G, HL, HM, K, L, M, P]

Ring-necked Duck, *Aythya collaris*
Very common spring and fall migrant and occasional non-breeding summer visitor throughout the region, to be expected from late March to May, and from mid-October to December. There are no non-historic (post-1920) breeding records for Nebraska (Ducey, 1988).
County Records: [A, B, C, D, F, G, HL, HM, K, L, M, P]

Green-winged Teal, adult pair

Greater Scaup, *Aythya marila*
Rare overwintering migrant, to be expected from late October to April. and from March 5 to May 10, including Rainwater Basin records (Jorgensen, 2012).
County Records: [A, B, C, HL, M, P]

Lesser Scaup, *Aythya affinis*
Very common spring and fall migrant throughout the region, with non-breeders occasionally present during summer. To be expected from mid-March to mid-October. Reported regionally for all months from October to May (Lingle, 1994), and from February 6 to August 7, including Rainwater Basin records (Jorgensen, 2012). Tier 2 At-Risk species.
County Records: [A, B, C, D, F, G, HL, HM, K, L, M, P]

Common Eider, *Somateria mollissima*
Very rare. A bird was shot in Lincoln County sometime between November 29 and December 2, 1967, and identified as *S. m. sedentaria* (NBR 37:38–39).
County Records: [L]

Harlequin Duck, *Histrionicus histrionicus*
Very rare. One individual was collected along the Platte River west of Grand Island (perhaps near Doniphan), Hall County about 1901 (Swenk, 1915b; Sharpe et al., 2001).
County Records: [HL]

Surf Scoter, *Melanitta perspicillata*
Rare. To be expected from late April to late May, and from October to December. There is an April regional record (Lingle, 1994). An immature or female was photographed at Harvard WPA, Clay County October 29, 2007 (Jorgensen, 2012).
County Records: [C, HL]

White-winged Scoter, *Melanitta fusca*
Rare migrant, to be expected from March to April, and from October to December. A bird was collected in Clay County February 14, 1916 (Sharpe et al., 2001; Jorgensen, 2012). There is also a specimen from Lincoln County for December 20, 1929 (Tout, 1947).
County Records: [A, B, C, G, L]

Black Scoter, *Melanitta americana*
Very rare migrant, to be expected from March to May, and from September to December. This is the rarest scoter species found in Nebraska, and like the others it is usually immature birds or females that are seen in the state.
County Records: [L]

Long-tailed Duck, *Clangula hyemalis*
Very rare migrant, to be expected from February to April, and from October to December.
County Records: [L]

Bufflehead, *Bucephala albeola*
Very common migrant throughout the region, to be expected from mid-March to late April, and from mid-October to late November
County Records: [A, B, C, D, F, G, HL, HM, K, L, M, P]

Common Goldeneye, *Bucephala clangula*
Very common migrant throughout the region, to be expected from early March to early April, and from mid-November to mid-December).
County Records: [A, B, C, D, F, G, HL, HM, K, L, M, P]

Bufflehead, adult pair

Barrow's Goldeneye, *Bucephala islandica*
Very rare spring and fall migrant. Species reported from North Platte, Lincoln County, January 23, 2008 and January 21–22, 2010, from Hall County November 25, 1995 (Sharpe et al., 2001) and from Buffalo County without a date or other attribution.
County Records: [B, HL, L]

Hooded Merganser, *Lophodytes cucullatus*
Common spring migrant, rare non-breeding summer visitor or resident, and occasional fall migrant, to be expected from late March to mid-November. Jorgensen noted that most Rainwater Basin summer records are for June, followed by July and August; the birds probably being non-breeding sub-adults. An early (1925) report of probable breeding at Harvard Marsh, Clay County, now seems unlikely to be credible (Jorgensen, 2012).
County Records: [A, B, C, F, G, HL, HM, K, L, M, P]

Common Merganser, *Mergus merganser*
Very common migrant throughout the region, to be expected from early March to late April, and from mid-November to mid-December. Occasionally overwinters on rivers, reservoirs or deeper wetlands, where it is rarely also a non-breeding summer visitor.
County Records: [A, B, C, D, F, G, HL, HM, K, L, M, P]

Red-breasted Merganser, *Mergus serrator*
Uncommon migrant, to be expected from late March to late April, and from early to late November.
County Records: [A, C, F, G, HL, M, P]

Ruddy Duck, *Oxyura jamaicensis*
Very common migrant throughout the region, and a highly local summer resident, to be expected from early April to mid-November. and reported to December 17, Adams County, by Jorgensen (2012). Broods were seen at Smith WPA, Clay County in 2001, and at Johnson WPA, Phelps County, in 1996 (Jorgensen, 2012).
County Records: [A, B, C, D, F, G, HL, HM, K, L, M, P]

Common Merganser, adult pair

Order Galliformes

Family Odontophoridae: New World Quail

Northern Bobwhite, *Colinus virginianus*
Local resident throughout the region.
County Records: [A, B, C, D, F, G, HL, HM, K, L, M, P]

Family Phasianidae: Pheasants, Grouse and Turkeys

[**Coturnix Quail**, *Coturnix coturnix*]
Hypothetical. Records are presumed to be the result of failed release efforts. After 1959 introductions in Frontier and Dawson counties, there have been no reports since 1962 (Shickley, 1968).
County Records: [D, F, L]

[**Chukar**, *Alectoris chukar*]
Hypothetical. The scattered regional records are presumed to be the result of failed release efforts. A pair was seen at Harvard WPA, Clay County, July 21, 2001 (NBR 69:114). Also reported from Lincoln County on April 8, 1974, April 14, 1977 and May 18, 1986 (Sharpe et al., 2001).
County Records: [A, L]

Gray Partridge, *Perdix perdix*
Very rare. A record exists for Hamilton County, July 21, 2005 (NBR 73:98). This species' current Nebraska range has greatly retracted, and it is now limited to a few northeastern counties.
County Records: [A, HM]

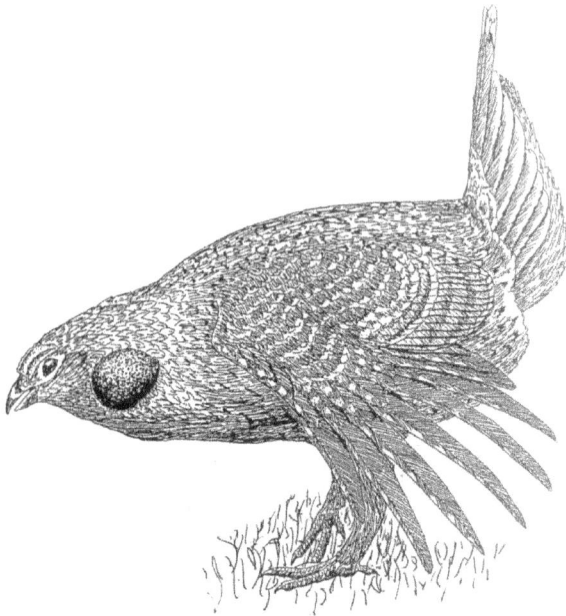

Sharp-tailed Grouse, male "dancing" display

Ring-necked Pheasant, *Phasianus colchicus*
Common resident throughout the region.
County Records: [A, B, C, D, F, G, HL, HM, K, L, M, P]

Sharp-tailed Grouse, *Tympanuchus phasianellus*
Rare local resident found in regional sandhills and sandy loess vegetation. Jorgensen (2012) listed a January 1976, record for Adams County as one of very few recent regional Rainwater Basin records. Mollhoff (2001) listed a probable breeding record for Hall County, and Tout (1947) regarded this species as a rare but increasing resident in Lincoln County.
County Records: [A, D, G, H, L]

Greater Prairie-Chicken, *Tympanuchus cupido*
Uncommon local resident throughout the region. Recently increasing in the Platte Valley and Rainwater Basin; and slowly increasing its range and abundance along the southern tier of central Nebraska

Greater Prairie-Chicken, adult male

counties. Display grounds (leks) were reported during 2011 on Shoe-maker Island, Hall County, and in several regional WPAs, including North Hultine in Clay County, Funk in Phelps County, and Clark, Lindau, Prairie Dog and Quadhammer, all in Kearney County. Displaying birds have also been seen Harvard, Hansen, Moger and Smith WPAs, all in Clay County (Jorgensen, 2012). Tier 1 At-Risk and ABC Red-List species.
County Records: [A, C, D, F, G, HL, K, L, M, P].

Wild Turkey, *Meleagris gallopavo*
Common reintroduced resident throughout the region. Since the re-introduction of this native but historically extirpated species starting in the 1960's, it has expanded its range along the central Platte Valley to encompass all of the regional counties.
County Records: [A, B, C, D, F, G, HL, HM, K, L, M, P]

Wild Turkey, adult male

Order Gaviiformes

Family Gaviidae: Loons

Red-throated Loon, *Gavia stellata.*
Rare migrant. Photographed during the fall of about 1899 in Frontier County (Sharpe et al., 2001) and on November 4, 2007 at Harvard WPA, Clay County (Jorgensen, 2012).
County Records: [C, F]

Common Loon, *Gavia immer*
Uncommon local migrant, to be expected from March to May, and from late October to early November. Jorgensen (2012) mentions seven Rainwater Basin reports from April 2 to May 15, two summer reports, and a fall report for Harvard Marsh, October 30, 1927.
County Records: [A, B, C, G, HL, P]

Order Podicipediformes

Family Podicipedidae: Grebes

Pied-billed Grebe, *Podilymbus podiceps*
Very common spring and fall migrant throughout the region, and a local summer resident. Jorgensen (2012) reported breeding at Moger WPA, Clay County in 2003, and found nests at Harvard WPA, Clay County in 2001. Evans and Wolfe (1967) found two broods in Clay County between 1958 and 1962.
County Records: [A, B, C, D, F, G, HL, HM, K, L, M, P]

Horned Grebe, *Podiceps auritus*
Very common spring and fall migrant, to be expected from mid-April to early May, and from early October to mid-November. There is an undocumented report of nesting in Lincoln County in 1963 (Sharpe et al., 2001). Jorgensen (2012) described this species as an uncommon spring migrant, and a rare to uncommon fall migrant, in the Rainwater Basin.
County Records: [A, B, C, D, F, HL, K, L, M, P]

Eared Grebe, *Podiceps nigricollis*
Very common spring and fall migrant throughout the region, and highly a local summer visitor or resident. Jorgensen (2012) consid-

ered this a fairy common spring and uncommon fall migrant, and a casual summer visitor or breeding resident in the Rainwater Basin. Nesting was observed at Harvard Marsh in 1914, in Clay County in 1984 and at Harvard WPA in 2007 (Mollhoff, 2001; Jorgensen, 2012). County Records: [A, B, C, D, F, G, HL, HM, K, L, M, P]

Western Grebe, *Aechmophorus occidentalis*
Very common spring and fall migrant and highly local summer visitor, to be expected from early May to early October. Jorgensen (2012) described this species as a casual spring and fall migrant and a summer visitor in the Rainwater Basin. There are records for Kissinger WPA, Clay County, and Funk WPA and Sacramento-Wilcox WMA, Phelps County. Tier 2 At-Risk species.
County Records: [B, C, D, F, G, HL, L, M, P]

Clark's Grebe, *Aechmophorus clarkii*
Rare spring and fall migrant, and a possible summer visitor. Similarity with Western Grebe makes determination of range limits difficult. Migration data are lacking but timing probably is comparable to the Western Grebe. Migrants or failed breeders have been seen during July at Sutherland Reservoir, Lincoln County (NBR 69: 109). Jorgensen (2012) reported a single bird at Funk WPA, Phelps County April 19, 1995. This hard-to-identify species may be more common regionally than is presently known. Previously considered a subspecies of the Western Grebe. Tier 2 Nebraska At-Risk and ABC Yellow-List species.
County Records: [F, HL, L, M, P]

Order Phaethontiformes

Family Phaethontidae: Tropicbirds

[White-tailed Tropicbird, *Phaethon lepturus*]
Hypothetical. A bird reported May 13–14, 1973 along the Platte River in Lincoln County may have actually been a tern with a fishing line attached (Bray et al., 1986; NBR 41:46; NBR 41: 79).

Top to bottom: American Bittern, Double-crested Cormorant,
and Pied-billed Grebe

Order Suliformes

Family Phalacrocoracidae: Cormorants

Neotropic Cormorant, *Phalacrocorax brasilianus*
Very rare migrant. Reported from Sutherland Reservoir, Lincoln County October 2, 1982 and May 2, 1999 (Wright, 1983; NBR 67:72–73). Also reported from Hultine WPA, Clay County May 31–June 3, 2008 (NBR 76: 98). This species is increasing in the southern Great Plains and is likely to be seen more frequently in the region.
County Records: [C, L]

Double-crested Cormorant, *Phalacrocorax auritus*
Very common spring and fall migrant throughout the region, and highly local summer resident, to be expected from mid-April to late September. Reported regionally for all months from March to December (Lingle, 1994).
County Records: [A, B, C, D, F, G, HL, HM, K, L, M, P]

Order Pelecaniformes

Family Pelecanidae: Pelicans

American White Pelican, *Pelecanus erythrorhynchos*
Very common spring and fall migrant throughout the region, and an uncommon non-breeding summer visitor, to be expected from late April to mid-October. Reported regionally for all months from March to December (Lingle, 1994).
County Records: [A, B, C, D, F, G, HL, HM, K, L, M, P]

Brown Pelican, *Pelecanus occidentalis*
Very rare. Reported from North Platte, Lincoln County, May 9, 1937 and May 6, 1952 (Sharpe et al., 2001).
County Records: [L]

Family Ardeidae: Bitterns and Herons

American Bittern, *Botaurus lentiginosus*
Uncommon spring and fall migrant throughout the region, and highly local summer resident, to be expected from early May to

American White Pelican, adult

early October. Reported regionally for all months from April to September, excepting June (Lingle, 1994). Jorgensen (2012) noted that, besides a historic record for Harvard Marsh, Clay County there were known nestings at Harvard WPA, Clay County, in 2007 and 2008, and a pair was seen there in 2001, There were also summer records for Adams and Clay counties during the 1970s, and a confirmed nesting in Clay County during the middle and late 1980s (Mollhoff, 2001).
County Records: [A, B, C, G, HL, HM, K, L, M, P]

Least Bittern, *Ixobrychus exilis*
Rare spring and fall migrant and highly local summer resident, to be expected from mid-May to mid-August. Reported regionally in May and August (Lingle, 1994). Swenk (1925) reported the species

to be a common summer resident and breeder at Inland Lagoon, Clay County in 1915–1916. Reported from Funk WPA July 24 to August 3, 1997 and June 12 to August 2 1998, and Kissinger Basin, Clay County June 6, 1999, Massie WPA in June 1981, and numerous summer sightings during recent years at Harvard and Hultine WPAs, Clay County (Sharpe et al., 2001; Jorgensen, 2012).
County Records: [A, B, C, HL, K, L, M, P]

Great Blue Heron, *Ardea herodias*
Very common spring and fall migrant throughout the region, and a local summer resident, to be expected from early April to mid-October. Reported regionally in all months (Lingle, 1994). There are nesting records for Clay, Dawson, Gosper, Hall, Lincoln and Phelps counties during the 1970's–1980's and probable nesting in Kearney County (Mollhoff, 2001; NBR 39:10–15). Jorgensen (2012) could found no evidence of recent nesting in the Rainwater Basin.
County Records: [A, B, C, D, F, G, HL, HM, K, L, M, P]

Great Egret, *Ardea alba*
Uncommon spring and fall migrant throughout the region, and a common non-breeding late-summer visitor. Reported regionally for all months from April to October (Lingle, 1994). This large egret has become increasingly common during recent decades, and especially so during post-breeding dispersal in August and September.
County Records: [A, B, C, D, F, G, HL, HM, K, L, M, P]

Snowy Egret, *Egretta thula*
Uncommon spring and fall migrant, regular late-summer visitor, and possible rare breeder. Reported regionally for all months from April to September (Lingle, 1994). Two adults with an immature bird seen in Hall County in late June during the 1980's offers evidence of regional breeding (Mollhoff, 2001).
County Records: [A, B, C, F, HL, HM, K, L, M, P]

Little Blue Heron, *Egretta caerulea*
Occasional spring and fall migrant and irregular summer visitor. Reported regionally for all months from April to October, excepting June (Lingle, 1994). Observed at Hultine WPA, Clay County, June 3, 2008 (NBR 78:99), and at North Hultine May 17, 1996. There are few June or July records, and most birds seen during early fall are white-plumaged juveniles.
County Records: [A, B, C, HL, K, L, M, P]

Tricolored Heron, *Egretta tricolor*
Rare summer and early fall visitor, to be expected mid-June through early September. Reported from Kearney County October 14, 1918 (Sharpe et al., 2001), Massie WPA, Clay County August 7, 1971 (Cink, 1973), Funk WPA, Phelps County, June 12, 2008 (NBR 76: 99), Funk WPA, Phelps County, July 27, 2003, and at a wetland in Clay County, August 30–September 2, 2003 (Jorgensen, 2012).
County Records: [C, K, P]

Reddish Egret, *Egretta rufescens*
Very rare. Reported from Funk WPA, Phelps County, June 12, 2008 (NBR 69:42–43). ABC Red-List species.
County Records: [P]

Cattle Egret, *Bubulcus ibis*
Uncommon spring and fall migrant and irregular summer visitor throughout the region. Reported regionally for all months from April to October, excepting July (Lingle, 1994). In the Rainwater Basin this species is most numerous in early fall (Jorgensen, 2012). There are no records of known breeding in this region or in the Rainwater Basin.
County Records: [A, B, C, D, G, HL, K, L, M, P]

Green Heron, *Butorides virescens*
Common spring and fall migrant and local summer resident or summer visitor throughout the region. Reported regionally for all months from April to October (Lingle, 1994). Evidence for current breeding in the Rainwater Basin is lacking (Jorgensen, 2012). Confirmed nesting for Dawson, Phelps, Hall and Lincoln counties during the mid- to latter 1980s was provided by Mollhoff (2001).
County Records: [A, B, C, D, F, HL, K, L, M, P]

Black-crowned Night-Heron, *Nycticorax nycticorax*
Common spring and fall migrant throughout the region. and local summer resident or summer visitor. Reported regionally for all months from April to October, excepting June (Lingle, 1994). Nesting in Lincoln County during the 1980's was reported by Mollhoff (2001). Breeding in the Rainwater Basin is local and irregular, with a colony reported at Hastings Basin, Adams County, in 1936, a nesting colony at Massie WPA, Clay County, in 1981, and a group of birds starting nests at Kissinger Basin, Clay County, in 2001 (Jorgensen, 2012). Tier 2 At-Risk species.
County Records: [A, B, C, D, F, G, HL, K, L, M, P]

Yellow-crowned Night-Heron, *Nyctanassa violacea*
Occasional spring and fall migrant, and summer visitor. Reported
regionally in May, July and August (Lingle, 1994). A juvenile seen
at Harvard WPA, Clay County, on June 24, 2001, suggested possible local breeding (NBR 69:112). Most of the regional summer reports are of juveniles or immatures that presumably were fledged
elsewhere.
County Records: [A, C, K, L, M, P]

Family Threskiornithidae: Ibises and Spoonbills

White Ibis, *Eudomimus albus*
Very rare summer visitor. Reported from Inland Lagoon, Phelps
County, June 12–19, 1916 (Sharpe et al., 2001), Harvard Marsh, Clay
County in June 1916, an immature seen at Kissinger Basin WPA,
Clay County, in July 1999, and an immature seen at Funk and John-
son WPAs, Phelps County, in August 2001 (Jorgensen, 2012).
County Records: [C, P]

White-faced Ibis, *Plegadis chihi*
Uncommon but increasingly frequent spring and fall migrant and
highly local summer resident. Reported regionally for all months
from April to November, excepting July (Lingle, 1994). White-faced
Ibises bred at Tauscher WPA, Fillmore County in 2005 (NBR 73:99),
and a nesting colony was found in 2007 at Harvard WPA, Clay
County (NBR 75:74). Nesting occurred at Harvard WPA in 2010
(NBR 78:91). Tier 2 At-Risk species.
County Records: [A, B, C, D, F, HL, HM, K, L]

Glossy Ibis, *Plegadis falcinellis*
Rare, but increasingly frequent spring and fall migrant and highly
local summer resident, to be expected from April to October. Two
adults were found among more than 70 White-faced Ibises at Harvard WPA, Clay County, in July 2001, and up to six were seen among
68 White-faced Ibises in early May 2006 (Jorgensen, 2012). From
2006 to 2011 there were six reports from Clay County, one from Adams County, and one from Hamilton County. Glossy Ibises were
also observed among nesting White-faced Ibises at Hultine WPA,
Clay County, in 2008 (NBR 76:99).
County Records: [A, C, HL, HM]

Roseate Spoonbill, *Platalea ajaja*
Very rare. One bird was seen June 5, 1932 in Buffalo County and two were seen near Hastings August 20, 1966 (Maunder, 1966; Sharpe et al., 2001). Two individuals were seen near Greenwood Cemetery, near Trumbull, Adams County, August 20, 1966 (NBR 4:66; Sharpe et al., 2001; Jorgensen, 2012).
County Records: [A, B, C]

Order Accipitriformes

Family Cathartidae: New World Vultures

[**Black Vulture**, *Coragyps atratus*]
Hypothetical. One bird was reportedly collected near Kearney, in October 1918 (Sharpe et al., 2001).
County Records: [B]

Turkey Vulture, *Cathartes aura*
Very common spring and fall migrant and highly local summer resident or visitor throughout the region. Reported regionally for all months from March to December, excepting November (Lingle, 1994). There are confirmed breeding records for Hall and Clay counties that date from the middle to late 1980's (Mollhoff, 2001).
County Records: [A, B, C, D, F, G, HL, HM, K, L, M, P]

Family Pandionidae: Ospreys

Osprey, *Pandion haliaetus*
Uncommon to occasional spring and fall migrant throughout the region. Reported regionally for all months from April to December, excepting July and November (Lingle, 1994).
County Records: [A, B, C, D, F, G, HL, HM, K, L, M, P]

Family Accipitridae: Kites, Hawks, and Eagles

Swallow-tailed Kite, *Elanoides forficatus*
Very rare. A specimen was collected at Harvard, Clay County, about 1910 (Sharpe et al., 2001). ABC Yellow-List species.
County Records: [C]

White-tailed Kite, *Elanus leucurus*
Rare migrant. One adult was seen Brady, Lincoln County, June 25, 2009 (NBR 77:99). Also reported July 1, 2008, Lincoln County (NBR 76:99). There are August to October regional records (Lingle, 1994). County Records: [HL, L]

Mississippi Kite. *Ictinia mississippiensis*
Rare spring and fall migrant and summer visitor, to be expected from mid-May to mid-September. Reported regionally for all months from May to August (Lingle, 1994). Tier 2 At-Risk species. County Records: [A, B, F, HL, L]

Bald Eagle, *Haliaeetus leucocephalus*
Very common overwintering migrant throughout the region, and a local breeding resident, to be expected from mid-November to late March. Reported regionally for all months (Lingle, 1994). Tier 2 At-Risk species.
County Records: [A, B, C, D, F, G, HL, HM, K, L, M, P]

Northern Harrier, *Circus cyaneus*
Uncommon spring and fall migrant throughout the region, and local summer resident, to be expected from mid-March to early December. Reported regionally for all months except June and July (Lingle, 1994). Jorgensen (2012) noted that there have been no nesting reports in the Rainwater Basin since the 1980's. In 1984 a nest was found at Harvard WPA, and Breeding Bird Atlas fieldwork produced confirmed breedings in Lincoln, Clay and Hall counties, as well as a probable breeding in Adams County (Mollhoff, 2001). County Records: [A, B, C, D, F, G, HL, HM, K, L, M, P]

Sharp-shinned Hawk, *Accipiter striatus*
Very common overwintering migrant throughout the region, to be expected from late November to late March. Reported regionally for all months except June and July (Lingle, 1994). Tier 2 At-Risk species.
County Records: [A, B, C, D, F, G, HL, HM, K, L, M, P]

Cooper's Hawk, *Accipiter cooperii*
Common overwintering migrant throughout the region, and local breeding resident, to be expected from mid-September to late April. Reported regionally for all months from September to April,

excepting February (Lingle, 1994). Breeding Bird Atlas fieldwork documented nesting in Adams County (Mollhoff, 2001). The species has recently increased in national and regional abundance, and increasingly has moved into suburban habitats for breeding. As a reflection of this trend, there were three active nests located near Hastings in 2010 (Jorgensen, 2012).
County Records: [A, B, C, D, F, G, HL, HM, K, L, M, P]

Northern Goshawk, *Accipiter gentilis*
Occasional to rare overwintering migrant, to be expected from September to late April. Reported regionally for all months from September to April, excepting February (Lingle, 1994). There are some early, 1896 and 1916, records from near Harvard, Clay County, and a 1957 record from Hamilton County (NBR 25:27; Sharpe et al., 2001), but no other Rainwater Basin records since 1974.
County Records: [A, B, C, F, HL, HM, L]

Red-shouldered Hawk, *Buteo lineatus*
Rare summer and fall visitor or migrant eastwardly, reported west to Lincoln County in the Platte Valley; there are scattered summer and fall records (June to October) from Adams, Clay, Hall, Lincoln and Phelps counties (Sharpe et al., 2012). This eastern *Buteo* is now nearly extirpated from eastern Nebraska. Tier 2 At-Risk species.
County Records: [A, C, HL, HM, L, M, P]

Broad-winged Hawk, *Buteo platypterus*
Rare spring and fall migrant eastwardly, to be expected from late April to mid-May and mid-September to early October. Reported regionally in April, May and August (Lingle, 1994). Nested in North Platte, Lincoln County during 2008 and 2009 (NBR 76:199; 77:99), and observed at Kearney, Buffalo County, Sept 22, 2010.
County Records: [A, B, L]

Swainson's Hawk, *Buteo swainsoni*
Very common spring and fall migrant throughout the region, and local summer resident, to be expected from mid-April to late September. Reported regionally for all months from April to September (Lingle, 1994). Breeding Bird Atlas confirmed nesting in Dawson, Hamilton and Lincoln counties, and probable nesting in Clay and Phelps counties (Mollhoff, 2001), Jorgensen (2012) listed a number of confirmed or probable nestings from the Rainwater Basin as recently

as the late 1990's, but noted there have been no recent summer reports from that region. Tier 2 At-Risk and ABC Yellow-List species.
County Records: [A, B, C, D, F, G, HL, HM, K, L, M, P]

Red-tailed Hawk, *Buteo jamaicensis*
Uncommon resident throughout the region. Reported regionally in
all months (Lingle, 1994). Breeding Bird Atlas fieldwork confirmed
nesting Adams, Clay, Dawson, Hall, Hamilton and Lincoln counties, and probable breedings in Buffalo, Clay, Frontier, Gosper, Merrick and Phelps counties (Mollhoff, 2001). Jorgensen (2012) reported
B. j. borealis as the local form in the Rainwater Basin, with *calurus,
krideri,* and *harlani* as seasonal migrants and/or winter visitors.
County Records: [A, B, C, D, F, G, HL, HM, K, L, M, P]

Red-tailed Hawk, adult

Top to bottom, left to right: Osprey, Northern Harrier, Black Tern,
Common Nighthawk, and Northern Saw-whet Owl

Ferruginous Hawk, *Buteo regalis*

Occasional to rare migrant throughout the region, rarely overwintering, and very rarely breeding in the west. Reported regionally in all months excepting April, May, June and July (Lingle, 1994). Nesting is mostly limited to the westernmost counties of Nebraska. Southward migration in winter occurs as prairie dogs and other rodents become less available. The dark-phase morph is rare in the region (Sharpe et al., 2001). Tier 1 At-Risk species.

County Records: [A, B, C, D, F, HL, K, L, M, P]

Rough-legged Hawk, *Buteo lagopus*
Common wintering migrant throughout the region, to be expected from early November to late March. Reported regionally for all months from October to April (Lingle, 1994).
County Records: [A, B, C, D, F, G, HL, HM, K, L, M, P]

Golden Eagle, *Aquila chrysaetos*
Rare fall and spring migrant, sometimes overwintering, and rare breeding resident in the west. Reported regionally for all months from October to April (Lingle, 1994). There are two breeding reports from Lincoln County in 1972 and 1973 (NBR 41:3–9; 43:13–19; Sharpe et al., 2001). Nesting is now mostly limited to the westernmost counties of Nebraska. Like the Ferruginous Hawk, Golden Eagles migrate southward during winter as prairie dogs become less available. Tier 2 At-Risk species.
County Records: [A, B, C, D, HL, L, M, P]

Order Falconiformes

Family Falconidae: Falcons

American Kestrel, *Falco sparverius*
Summer resident or spring and fall migrant throughout the region, with lower winter populations. Reported in all months (Lingle, 1994). Breeding Bird Atlas fieldwork confirmed nesting in Adams, Clay, Dawson, Gosper, Hall, Hamilton, Lincoln and Merrick counties, and probable nesting in Buffalo, Frontier, Kearney and Phelps counties (Mollhoff, 2001).
County Records: [A, B, C, D, F, G, HL, HM, K, L, M, P]

Merlin, *Falco columbarius*
Very common overwintering migrant throughout the region, to be expected from October to March. Reported regionally for all months from November to April (Lingle, 1994). Jorgensen (2012) indicated that fall and spring records are likely to represent *F. c. columbarus,* whereas *F. c. richardsonii* has been seen from November to February. There is no evidence of regional or state nesting. Tier 2 At-Risk species.
County Records: [A, B, C, D, F, G, HL, HM, K, L, M, P]

Gyrfalcon, *Falco rusticus*
Very rare wintering migrant, to be expected from November to March. One was observed near Aurora, Hamilton County, December 31, 2011 by Johnsgard and three others, one seen in Hall County October 20, 1978 (Sharpe et al., 2001), and an immature was captured by a falconer west of Minden, Kearney County on December 14, 1974 (NBR 44:3). Gyrfalcons are regularly seen and sometimes captured by falconers in the Nebraska Sandhills during winter (R. Berry, pers. comm.).,
County Records: [HL, HM, K]

Peregrine Falcon, *Falco peregrinus*
Increasingly common and sometimes overwintering spring and fall migrant, to be expected from mid-September to late March. Reported regionally from February to May and in August, October and December (Lingle, 1994). No current or documented historic Nebraska nesting records exist other than urban-adapted birds nesting in Lincoln and Omaha. Tier 2 At-Risk species.
County Records: [A, B, C, D, HL, K, L, M, P]

Prairie Falcon, *Falco mexicanus*
Uncommon overwintering migrant, especially westward. Reported regionally for all months from September to March (Lingle, 1994). There are no regional nesting records. Tier 2 Nebraska At-Risk species.
County Records: [A, B, C, D, G, K, L, M, P]

Order Gruiformes

Family Rallidae: Rails, Gallinules, and Coots

Yellow Rail, *Coturnicops noveboracensis*
Very rare spring and fall migrant. A specimen in the Hastings museum was collected at Harvard Marsh, Clay County June 12, 1920 (Jorgensen, 2012). ABC Red-List species.
County Records: [C, HL, HM, P]

Black Rail, *Laterallus jamaicensis*
Very rare spring and fall migrant. Seen at Funk Lagoon, Phelps County May 13, 1979 (Sharpe et al., 2001). ABC Red-List species.
County Records: [P]

King Rail, *Rallus elegans*
Very rare spring and fall migrant and historic summer resident (east), to be expected from May to August. The species is rare and mostly limited to a few wetlands in eastern Nebraska. There was specimen collected at Harvard Marsh, Clay County in June 1919, another was adult taken there in September 1922, and juveniles were collected at Harvard Marsh, Clay County and at Inland, Clay County in October 1916 (Jorgensen, 2012). Tier 1 At-Risk and ABC Yellow-List species.
County Records: [A, HM]

Virginia Rail, *Rallus limicola*
Uncommon to rare spring and fall migrant and a highly local summer resident, to be expected from early May to mid-September. Reported regionally for all months from April to October (Lingle, 1994). Jorgensen (2012) listed some nesting records for the eastern Rainwater Basin. Two adults with four chicks were seen at Funk Lagoon, Phelps County July 4, 1998 (Sharpe et al., 2001), birds were also seen there July 9, 1995, May 17–19, 1996 and June 30, 1996 (Sharpe et al., 2001).
County Records: [A, B, C, D, HL, HM, P]

Sora, *Porzana carolina*
Common spring and fall migrant and a probable local or rare summer resident, to be expected from early May to late September. Reported regionally for all months from April to October, except for July (Lingle, 1994). Jorgensen (2012) noted there is an influx of Soras into the Rainwater Basin during July, possibly being migrants rather than breeding birds.
County Records: [A, B, C, D, F, G, HL, HM, K, L, P]

Common Moorhen, *Gallinula galeata*
Rare to uncommon spring and fall migrant and a probable rare summer resident, to be expected from mid-May to late August. A juvenile was seen in Clay County, August 30, 1998, providing the first regional nesting record (NBR 69:115). Jorgensen (2012) provided breeding evidence only for the eastern Rainwater Basin (Fillmore and Seward counties).
County Records: [C, L]

American Coot, *Fulica americana*
Very common spring and fall migrant throughout the region and local summer resident. It may be expected from late March to early

November, but may remain into early winter on ice-free waters. Reported regionally for all months (Lingle, 1994). Jorgensen (2012) reported breeding records for Harvard, Hultine and Moger WPAs, all in Clay County.
County Records: [A, B, C, D, F, G, HL, HM, K, L, M, P]

Family Gruidae: Cranes

Sandhill Crane, *Grus canadensis*
Very common spring and fall migrant, to be expected from early March to mid-April, and from early October to early November) and very rare summer regional resident. Reported regionally for all months; several thousand birds overwintered between Kearney and Grand Island during the winter of 2011–2012. Records of recent nesting in the Rainwater Basin have become more common since the early 1990's (Jorgensen, 2012; Johnsgard, 2012b). Tier-I At-Risk species.
County Records: [A, B, C, D, F, G, HL, HM, K, L, M, P]

Sandhill Crane, adult

Common Crane, *Grus grus*
Very rare spring migrant. This Eurasian species was observed in the
region every spring from 2007 to 2012. One appeared with early-mi-
grating Sandhill Cranes on January 27, 2012, near Alda Road, Hall
County. It remained in that general area with Sandhill Cranes until
at least February 27.
County Records: [A, B, HL, L, P]

Whooping Crane, *Grus americana*
Very rare spring and fall migrant, to be expected from late March
to early May, and from mid-September to early November. In 2012,
the Central Flyway population consisted of less than 300 individuals, Reported regionally by Lingle (1994) from March to May and
from October to December). On January 21, 2012, an adult Whoop-

Whooping Crane, adult

ing Crane appeared near Alda, Hall County among overwintering Sandhill Cranes. This crane, or a similarly lone Whooping Crane, has appeared every year in the Alda area with Sandhill Canes since at least 2007, and in 2012 it remained until at least March 18. Tier 1 At-Risk, ABC Red-List, federal ESA endangered, and NESCA endangered species.
County Records: [A, B, C, D, F, G, HL, HM, K, L, M, P]

Hooded Crane, *Grus monacha*
Very rare spring migrant. This east Asian crane was reported from March 25 to April 11, 2011, in Hall County (NBR 79:53), and was seen for several days from January 27, 2012, near Alda, Hall County. It has been speculated that all North American Hooded Crane records reflect escapes from captivity, but in 2012 a Hooded Crane, perhaps this same bird, appeared in February, at Hiwassee NWR, Tennessee, and near Greene, Indiana.
County Records: [HL]

Order Charadriiformes

Family Charadriidae: Plovers

Black-bellied Plover, *Pluvialis squatarola*
Uncommon to fairly common spring migrant throughout region, to be expected from early April to late June; rare fall migrant, and uncommon summer visitor most likely to be seen late July to mid-November. Generally more juveniles than adults during fall migration (Sharpe et al., 2001). USSCP species of moderate concern.
County Records: [A, C, D, G, HL, HM, K, L, P]

American Golden-Plover, *Pluvialis dominica*
Fairly common spring and fall migrant, rare summer visitor, migrants to be expected from early April to mid-May, and mid-August to mid-November. Spring migration passes through the interior of North American and most fall migration passes off the Atlantic Coast. Occurs most often in the eastern Rainwater Basin, rarely seen west of Phelps County (Sharpe et al., 2001). Any *Pluvialis* plovers seen in the fall might be Pacific Golden-Plovers, so pertinent field marks should be closely scrutinized (Sharpe et al., 2001). USSCP species of high concern, CSCP priority and ABC Yellow-List species.
County Records: [A, C, G, HL, HM, K, L, P]

Snowy Plover, *Charadrius nivosus*
Rare spring and fall migrant, and very rare local breeder; migrants
to be expected from late April to late May and in August. May be
confused with Piping and Semipalmated plovers, so pertinent field
marks should be closely scrutinized. ABC Yellow-List species.
County Records: [A, B, C, L, P]

Mountain Plover, *Charadrius montana*
Very rare migrant, to be expected from May through July. Tout
(1947) reported the species may have nested in Lincoln County in
1859. ABC Red-List species, Tier 1 At-Risk and NESCA threatened
species and candidate for federal ESA threatened status.
County Records: [L]

Semipalmated Plover, *Charadrius semipalmatus*
Fairly common spring and fall migrant throughout region, to be ex-
pected from mid-April to mid-June, and late July to mid-November.
May be confused with Snowy or Piping plovers so pertinent field
marks should be closely scrutinized. USSCP species of low concern.
County Records: [A, B, C, D, F, G, HL, K, L, P]

Piping Plover, *Charadrius melodus*
Rare spring and fall migrant and regular, rare, local breeder, to be
expected from mid-April to late August. Population size increasing
as the result of extensive conservation efforts in the region. Nesting habitat occurs along river systems and includes open, bare
sand found on sandbars, sand and gravel mines and dredging operations. May be confused with Snowy or Semipalmated plovers,
so pertinent field marks should be closely scrutinized. Federal
ESA threatened, NESCA threatened, ABC Red-List, Tier 1 At-Risk,
USSCP highly imperiled and CSCP priority species.
County Records: [A, B, C, D, HL, K. L, P]

Killdeer, *Charadrius vociferus*
Abundant resident throughout region, seen in all months; and
common spring and fall migrant and regular breeder, migrants to
be expected from early March to early November. Typically found
in wetlands, agricultural fields, hayfields, sod farms, pastures,
grassland burns, planted or fallow farm fields and short-grass prairie. USSCP species of moderate concern.
County Records: [A, B, C, D, F, G, HL, HM, K, L, M, P]

Family Recurvirostridae: Stilts and Avocets

Black-necked Stilt, *Himantopus mexicanus*
Casual spring and rare fall migrant, rare summer resident and breeder, migrants to be expected from mid-April to late May, and late July to mid-September. Sightings have become more frequent since 1997. (Jorgensen, 2012). Before 2008, the species had not been recorded during August and September. USSCP species of low concern.
County Records: [A, C, D, HL, L, P]

American Avocet, *Recurvirostra americana*
Fairly common spring and uncommon fall migrant throughout region, and rare breeder, migrants to be expected from mid-April to early June, and late June to early October. Nests found in Clay County, and young birds seen in Phelps County in 1974 (Bennett, 1975; Mollhoff, 2001). Tier 2 At-Risk species, USSCP species of moderate concern, CSCP priority species.
County Records: [A, B, C, D, F, G, HL, K, L, M, P]

Family Scolopacidae: Sandpipers and Phalaropes

Spotted Sandpiper, *Tringa macularia*
Fairly common spring and fall migrant throughout region, and rare, regular breeder, migrants to be expected from late April to early June, and late August to mid-September. Easily identified by distinctive wing-beat and bobbing pattern. USSCP species of moderate concern.
County Records: [A, B, C, D, F, G, HL, HM, K, L, M, P]

Solitary Sandpiper, *Tringa solitaria*
Uncommon spring migrant and fairly common fall migrant and rare summer visitor, migrants to be expected from late April to early June and late June to early September. Both *T. s. solitaria* and *T. s. cinnamomea* occur in Nebraska (Sharpe et al., 2001). USSCP species of high concern.
County Records: [A, B, C, D, F, G, HL, HM, K, L, M, P]

Greater Yellowlegs, *Tringa melanoleuca*
Fairly common spring and fall migrant throughout region, and ca-

sual winter visitor, migrants to be expected from mid-March to mid-June, and late June to late November. More Lesser than Greater Yellowlegs are usually seen in the region. USSCP species of moderate concern.
County Records: [A, B, C, D, F, G, HL, HM, K, L, M, P]

Willet, *Tringa semipalmata*
Fairly common spring and rare fall migrant and uncommon summer resident, migrants to be expected from mid-April to late May, and mid-July (adults) to late August (juveniles). Migratory populations of *T. s. inornatus* are present in the region. Tier 2 At-Risk species and USSCP species of moderate concern.
County Records: [A, B, C, F, G, HL, K, L, P]

Lesser Yellowlegs, *Tringa flavipes*
Common to abundant spring and common fall migrant throughout the region, to be expected from mid-March to early June, and late June to late October, with some non-breeding summer visitors. USSCP species of moderate concern.
County Records: [A, B, C, D, F, G, HL, HM, K, L, M, P]

Lesser Yellowlegs, adult

Upland Sandpiper, *Bartramia longicauda*
    Fairly common spring and fall migrant throughout the region, and
    regular breeder, to be expected from early April to mid-September.
    Population decline was caused by market hunting and by habitat
    alteration, reducing the amount of suitable nesting habitat (Sharpe
    et al., 2001). USSCP species of high concern and CSCP priority
    species.
    County Records: [A, B, C, D, F, G, HL, HM, K, L, M, P]

Eskimo Curlew, *Numenius borealis*
    Almost certainly extinct; formerly a common spring and uncommon
    fall migrant, to have been expected from April to June, and October
    (Sharpe et al., 2001; Jorgensen, 2012). Eskimo Curlews were proba-
    bly most frequent in the eastern sections of the region (Jorgensen,
    2012). Swenk (1915) noted that "the chief feeding grounds of these

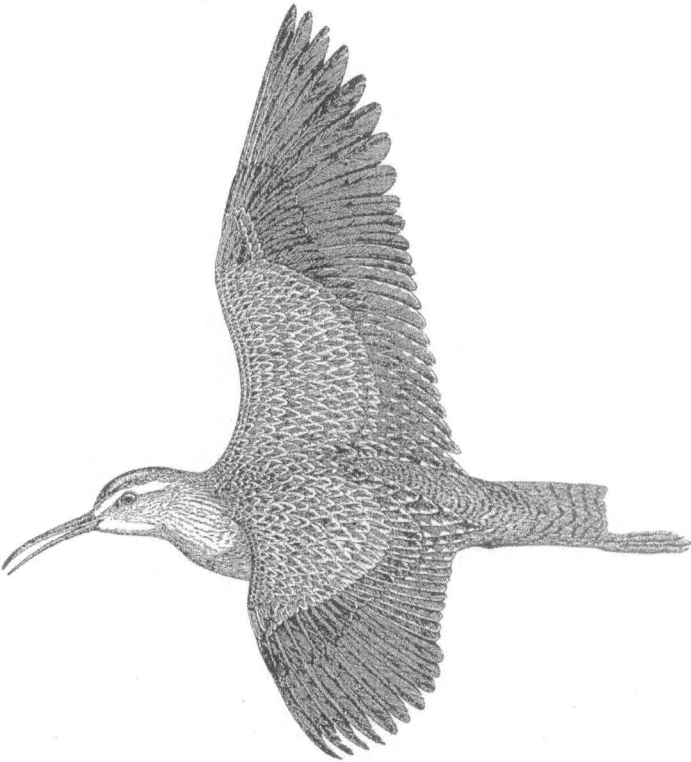

Eskimo Curlew, adult

curlews at the time (1877) were in York, Fillmore, and Hamilton counties, and their heaviest lines of northward migration between the 97th and 98th meridian." In the fall of 1880, Townsley noted that the birds were "common during the fall migration" (Swenk, 1925). Townsley also noted in April 1887 that the "birds were becoming so scarce he thought he had better add a pair to his collection before it was too late" (Brooking, 1942). The last record for Nebraska was on April 8, 1926 five miles east of Hastings, Clay County (Brooking, 1942). Federal ESA endangered, NESCA endangered, ABC Red-List and USSCP highly imperiled (probably extinct) species.
County Records: [A, C, HM, M]

Whimbrel, *Numenius phaeopus*
Rare spring and fall migrant, to be expected from late April to mid-May, and late July to mid-October. Most observations have occurred since 1995 (Jorgensen, 2012). USSCP species of high concern.
County Records: [A, B, C, HL, L]

Long-billed Curlew, *Numenius americanus*
Occasional spring and rare fall migrant, to be expected from late

Long-billed Curlew, adult female

March to mid-April. and September. Breeds regularly in the western Nebraska Sandhills. Swenk (1925) reported finding individuals near Funk, Phelps County in the 1890's. Tier 1 At-Risk species, USSCP species of high concern, CSCP priority species and ABC Yellow-List species.
County Records: [B, F, HL, L, P]

Hudsonian Godwit, *Limosa haemastica*
Uncommon spring and fall migrant, to be expected from early April to mid-June, and late August to early September. ABC Yellow-List species.
County Records: [A, B, C, HL, K, L, P]

Marbled Godwit, *Limosa fedoa*
Uncommon spring and casual fall migrant, to be expected from late March to late May, and mid-June to mid-September. USSCP species of high concern and ABC Yellow-List species.
County Records: [A, B, C, F, G, HL, L, P]

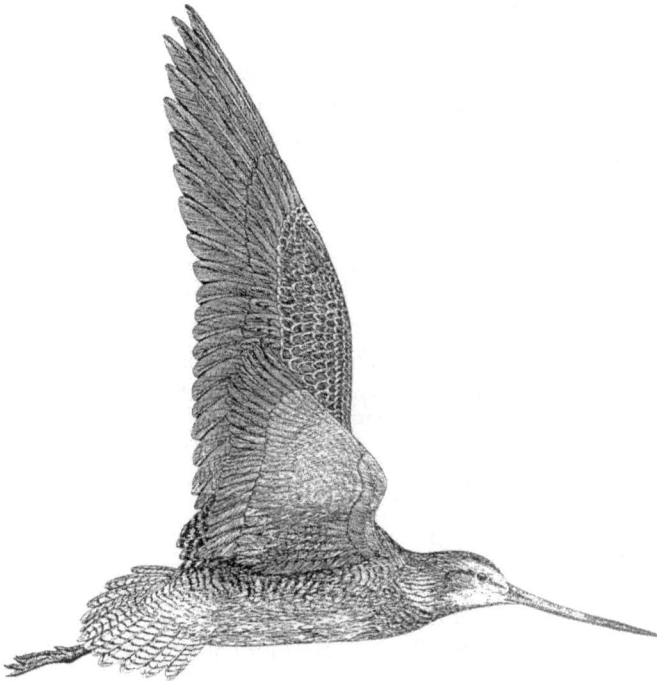

Marbled Godwit, adult

Ruddy Turnstone, *Arenaria interpres*
Occasional spring and rare fall migrant, most likely to be se to be expected from early May to early June, and late July to early September. USSCP species of high concern.
County Records: [A, B, C, D, HL, L, P]

Red Knot, *Calidris canutus*
Very rare spring and fall migrant, to be expected from late April to late May, and mid-August to mid-September. USSCP species of high concern and ABC Yellow-List species.
County Records: [A, C, HL, L, P]

Sanderling, *Calidris alba*
Uncommon spring and occasional fall migrant, to be expected from late March to early June, and early July to mid-November. In fall, adults migrate earlier than juveniles. Spring and fall migratory stopover increased after large reservoirs filled in 1930's and 1940's (Tout, 1947). USSCP species of high concern and ABC Yellow-List species.
County Records: [A, C, G, HL, K, L, P]

Semipalmated Sandpiper, *Calidris pusilla*
Abundant spring and fairly common fall migrant throughout the region, to be expected from mid-April to mid-June, and early July to mid-October. Often migrates with White-rumped, Least and Baird's Sandpipers. USSCP species of moderate concern, CSCP priority species, and ABC Yellow-List species.
County Records: [A, B, C, D, F, G, HL, HM, K, L, M, P]

Western Sandpiper, *Calidris mauri*
Occasional spring and uncommon fall migrant, mainly in western parts of region, to be expected from mid-April to late May, and early July to late October. USSCP species of high concern and ABC Yellow-List species.
County Records: [A, C, D, F, HL, K, L, P]

Least Sandpiper, *Calidris minutilla*
Common spring and fairly common fall migrant throughout the region, to be expected from late March to mid-June, and early June to early November. USSCP species of moderate concern and CSCP priority species.
County Records: [A, B, C, D, F, G, HL, HM, K, L, M, P]

White-rumped Sandpiper, *Calidris fuscicollis*
Abundant spring migrant, very rare fall migrant throughout the region, to be expected from mid-April to late June, and late July. Spring migration occurs through interior North America, and fall migration off the Atlantic Coast. Often seen with Semipalmated, Least and Baird's sandpipers during migratory stopovers. USSCP species of low concern, CSCP priority species and ABC Yellow-List species.
County Records: [A, B, C, D, F, G, HL, HM, K, L, M, P]

Baird's Sandpiper, *Calidris bairdii*
Common spring and uncommon fall migrant throughout the region, to be expected from mid-March to mid-June, and late June to early November. Tends to arrive earlier in the region than other small *Calidris* species. USSCP species of low concern and CSCP priority species.
County Records: [A, B, C, D, F, G, HL, HM, K, L, M, P]

Pectoral Sandpiper, *Calidris melanotos*
Common spring and fall migrant throughout the region, to be expected from mid-March to mid-June, and late June to early November. USSCP species of low concern and CSCP priority species.
County Records: [A, B, C, D, F, G, HL, HM, K, L, M, P]

Sharp-tailed Sandpiper, *Calidris acuminata*
Very rare fall migrant, to be expected from late September to mid-October. All records are of juvenile birds (Jorgensen, 2012).
County Records: [G, P]

Dunlin, *Calidris alpina*
Uncommon late spring and rare fall migrant, to be expected from early April to early June, and early October to early November. Adults and juveniles tend to migrate together (Sharpe et al., 2001). USSCP species of high concern.
County Records: [A, C, D, HL]

Curlew Sandpiper, *Calidris ferruginea*
Very rare summer visitor; one bird seen at Funk SPA on July 19–21, 1997 (Jorgensen and Silcock, 1998; Brogie, 1998; NBR 65: 107).
County Records: [P]

Stilt Sandpiper, *Calidris himantopus*
Common to abundant spring and common fall migrant throughout the region, to be expected from mid-April to early June, and early July to late October. USSCP species of moderate concern, CSCP priority species and ABC Yellow-List species.
County Records: [A, B, C, D, F, G, HL, HM, K, L, P]

Buff-breasted Sandpiper, *Tryngites subruficollis*
Fairly common spring and uncommon fall migrant, to be expected from late April to early June, and late June to late September. Spring and fall migration pass through interior North America; fall reports in the region are likely of juveniles (Sharpe et al., 2001). The eastern Rainwater Basin is a critically important stopover location during spring migration (Skagen et al., 1999; Sharpe et al., 2001; Jorgensen 2004, 2012). Most often seen in agricultural (soybean) fields (Jorgensen, 2012). Tier 1 At-Risk species, USSCP species of high concern, CSCP priority species and ABC Red-List species.
County Records: [A, C, D, K, L, P]

Ruff, *Philomachus pugnax*
Very rare spring and fall migrant, to be expected from late March to late May, and late July to late September. A juvenile was in a flooded field near Axtell, Phelps County on September 22–23, 1993, a red-morph male was at Sacramento-Wilcox WMA, Phelps County on April 19, 1994, a black-morph male was at Funk WPA, Phelps County on May 24, 1997 and a male in basic plumage was seen at Eckhardt WPA, Clay County on March 26, 2005 (Jorgensen, 1994b, 2012; Sharpe et al., 2001; NBR 65:82).
County Records: [C, K, P]

Short-billed Dowitcher, *Limnodromus griseus*
Uncommon spring and fall migrant, to be expected from late April to late May, and early July to mid-September. Adult migrants are more common in July and juvenile migrants more common in August and September. *L. g. hendersoni* occurs in region (Jorgensen, 1996, 2012). USSCP species of high concern.
County Records: [A, B, C, D, F, HL, HM, L, P]

Long-billed Dowitcher, *Limnodromus scolopaceus*
Abundant spring and common fall migrant throughout the region, to be expected from mid-March to early June, and mid-July to mid-

November. Uncommon in the region before large reservoirs completed in the 1930's and 1940's (Tout, 1947). USSCP species of low concern and CSCP priority species.
County Records: [A, B, C, F, G, HL, HM, K, L, M, P]

Wilson's Snipe, *Gallinago delicata*
Fairly common spring and fall migrant, rare, local breeder and winter visitor, to be expected from mid-March to early May, and late June to early November. Tier 2 At-Risk species and USSCP species of moderate concern.
County Records: [A, B, C, D, G, HL, HM, K, L, M, P]

American Woodcock, *Scolopax minor*
Uncommon spring and fall migrant and regular breeder, to be expected from mid-March to early October. More common in east, but moving westward as suitable riparian woodland habitat expands; nests as far west as Kearney County (Sharpe et al., 2001). Tier 2 At-Risk species and USSCP species of high concern.
County Records: [B, C, HL, K, L, M]

Wilson's Phalarope, *Phalaropus tricolor*
Common to abundant spring and uncommon to fairly common fall migrant, rare summer visitor and casual breeder, to be expected from mid-April to late May, and early July to late October. Polyandrous, regularly nests in the Sandhills northwest of the central Platte Valley USSCP species of high concern.
County Records: [A, C, D, F, HL, K, L, M, P]

Red-necked Phalarope, *Phalaropus lobatus*
Very rare spring and occasional fall migrant, to be expected from mid-April to late May and early August to mid-October. USSCP species of moderate concern.
County Records: [A, C, D, K, L, P]

Red Phalarope, *Phalaropus fulicaria*
Very rare spring and summer migrant, to be expected from late April to mid-May and in August. A female in worn breeding plumage was seen in Phelps County on August 1, 1993, a female in basic plumage was seen at Funk WPA, Phelps County on August 24, 1993 and a bird in basic plumage was seen at North Hultine WPA, Clay County on June 15, 2001 (Silcock, 2001; Jorgensen, 2012). USSCP species of moderate concern.
County Records: [C, G, K, P]

Family Laridae: Gulls and Terns

[Jaeger species, *Stercorarius* spp.]
Rare late summer and fall migrants, to be expected from early July to mid-November. Most reports are probably of juvenile Pomarine or Parasitic jaegers (Sharpe et al., 2001). Reports come from North Platte, Lincoln County in November 1885, Kearney County in November 1919, October and November 1926, Adams County in October 1927 or 1928 and Dawson County in July 1975 (Robertson, 1977; Sharpe et al., 2001; Jorgensen, 2012).
County Records: [A, D, L, K]

Pomarine Jaeger, *Stercorarius pomarinus*
Very rare fall migrant, to be expected from October to November. Reported from October 1917 in Buffalo County and Lincoln County in November 1895 (Bruner, 1896; Swenk, 1915a; Tout, 1947; Sharpe et al., 2001).
County Records: [B, L]

Black-legged Kittiwake, *Rissa tridactyla*
Very rare winter visitor, to be expected from late November to late December. Typically seen at large lakes and reservoirs, such as Johnson Lake, Gosper County on December 23, 1991.
County Records: [G, HL]

Sabine's Gull, *Xema sabini*
Very rare fall and winter visitor, to be expected from late September to late November. Seen at large lakes and reservoirs, such as Medicine Creek Reservoir, Frontier County on September 27, 2010 and Lake Maloney on September 26 1996 in Lincoln County.
County Records: [F, L]

Bonaparte's Gull, *Coriocephalus philadelphia*

Uncommon migrant and occasional local summer visitor throughout the region, to be expected from early April to early June, and early September to late December. Usually seen at large lakes and reservoirs during stormy weather years, but was also seen at Harvard Marsh, Clay County in 1915 (Sharpe et al., 2001).
County Records: [A, B, C, D, F, G, HL, K, L, M, P]

Little Gull, *Hydrocoleus minutus*
Very rare spring and fall visitor. A first-year bird was seen at Massie WPA, Clay County on May 6, 2000 (Jorgensen, 2002, 2012), and another seen at Sutherland Reservoir, Lincoln County on November 1, 1998 (Sharpe et al., 2001).
County Records: [C, L]

Ross's Gull, *Rhodostethia rosea*
Very rare winter visitor. One bird was seen at Sutherland Reservoir in Lincoln County on 17–23 December 1992 (Dinsmore and Silcock, 1993; Sharpe et al., 2001). ABC Yellow-List species.
County Records: [L]

Laughing Gull. *Leucophaeus atricilla*
Rare spring and fall migrant and summer visitor, to be expected from late March to late May, and late October to late December. Most reports are of juveniles or birds in first or second basic or second alternate plumage (Sharpe et al., 2001).
County Records: [C, HL, L, P]

Franklin's Gull, *Leucophaeus pipixcan*
Very common spring and fall migrant throughout the region, to be expected from late February to May, and from September to late November, A rare summer visitor; such visitors tend to be first-year birds (Sharpe et al., 2001). This is the most common black-headed gull in the region.
County Records: [A, B, C, D, F, G, HL, HM, K, L, M, P]

Mew Gull, *Larus canus*
Very rare spring visitor. One bird seen at Johnson Lake, Gosper County on February 21, 1999. *L. c. brachyrhynchus* occurs in the region (Sharpe et al., 2001).
County Records: [G, L]

Ring-billed Gull, *Larus delawarensis*
Abundant spring and fall migrant and common summer and winter resident, with reports from all months, to be expected from mid-February to late March and early August to late November. Most birds observed in the summer are immatures, and the number of birds seen in winter varies with the severity of the weather (Sharpe et al., 2001).
County Records: [A, B, C, D, F, G, HL, HM, K, L, M, P]

California Gull, *Larus californicus*
Occasional spring and fall migrant and rare winter visitor, to be expected from late February to late May and mid-July to late December. Numbers have increased over the past 50 years, with most observations at large lakes and reservoirs such as Lake Maloney and Sutherland Reservoir. *L. c. albertaensis* and *L. c. californicus* are both possible in region (Sharpe et al., 2001).
County Records: [G, K, L]

Herring Gull, *Larus argentatus*
Occasional spring and fall migrant, to be expected from late February to late May and early August to late October.
County Records: [A, B, D, F, G, HL, K, L, P]

Thayer's Gull, *Larus thayeri*
Occasional spring and fall migrant and winter visitor, most likely to be expected from late February to late April, and early November to late January. This arctic-breeding species is probably more common in region, but may be overlooked due to similarity to Herring Gull. ABC Yellow-List species.
County Records: [L]

Lesser Black-backed Gull, *Larus fuscus*
Very rare winter visitor. Seen at Sutherland Reservoir, Lincoln County and Johnson Lake in Gosper County.
County Records: [G, L]

Glaucous Gull, *Larus hyperboreus*
Very rare winter visitor. Single birds observed in Holdrege, Phelps County on December 8–11 2008, North Platte, Lincoln County on December 18, 2010 and at Sutherland Reservoir, Lincoln County on January 1–3, 2011; also reported from Elwood Reservoir, Dawson County and Johnson Lake, Lincoln County (Sharpe et al., 2001).
County Records: [D, G, L, P]

Great Black-backed Gull, *Larus marinus*
Very rare winter visitor, to be expected from January to February. Observed at Johnson Lake, Gosper County on February 21, 1999 and at Sutherland Reservoir, Lincoln County on January 14, 1989 (Sharpe et al., 2001).
County Records: [G, L]

Least Tern, *Sterna antillarum*
Rare spring and fall migrant, to be expected from early May through mid-August; a rare but regular summer breeder. Population is increasing in the region as the result of extensive regional conservation efforts. Nesting habitat occurs along river systems and includes open, bare sand found on sandbars, sand and gravel mines, and dredging operations. Interior race, *S. a. athalassos* occurs in the region. Federal ESA endangered species, NESCA endangered species, Tier 1 At-Risk species and ABA Red-List species.
County Records: [A, B, D, HL, HM, K, L, M, P]

Caspian Tern, *Sterna caspia*
Rare spring and fall migrant, to be expected from late March to early June and late July to early October.
County Records: [F, G, HL, L, P]

Black Tern, *Chlidonias niger*
Very common spring and fall migrant throughout the region, and rare, local breeder, migrants to be expected from mid-April to early October. Nests in the Sandhills northwest of the region but may breed in the Rainwater Basin region when the basins hold water (Bruner et al., 1904; Sharpe et al., 2001; Jorgensen, 2012). Tier 2 At-Risk species.
County Records: [A, B, C, D, F, G, HL, HM, K, L, M, P]

Common Tern, *Sterna hirundo*
Rare spring and fall migrant, to be expected from mid-May to mid-June, and late July to early October.
County Records: [C, HL, L, P]

Forster's Tern, *Sterna forsteri*
Abundant spring and fall migrant throughout the region. and rare, local breeder, migrants to be expected from late March to mid-June, and mid-July to early November. Tier 2 At-Risk species.
County Records: [A, B, C, D, F, G, HL, HM, K, L, M, P]

Order Columbiformes

Family Columbidae: Pigeons and Doves

Rock Pigeon, *Columba livia*
Abundant permanent resident throughout the region, and regular breeder in all seasons but winter. Introduced, but not clear when the species arrived in Nebraska. Swenk (1918) did not mention species, but Haecker et al. (1945) reported it as being present statewide. May be declining in Nebraska due to eradication programs and loss of nesting habitat. Competition with Eurasian Collared-Doves may accelerate the decline.
County Records: [A, B, C, D, F, G, HL, HM, K, L, M, P]

Eurasian Collared-Dove, *Streptopelia decaocto*
Common permanent resident and regular breeder throughout the region. Range is rapidly expanding and numbers increasing. Introduced into the U.S. from the Old World via Florida in the mid-20th century; first observed in the Platte Valley in Kearney, Buffalo County on November 29, 1997. Also seen in Kearney, Buffalo County on April 22, 1998 and in Shelton, Buffalo County on May 31, 1998 (Brogie, 1998; Sharpe et al., 2001). Nesting in the state began by 1998, and by 2012 it had been reported from all of Nebraska's 93 counties.
County Records: [A, B, C, D, F, G, HL, HM, K, L, M, P]

White-winged Dove, *Zenaida asiatica*
Rare spring and very rare summer, fall, winter visitor, to be expected from mid-May to early July. Seen with Eurasian Collared-Doves in Kearney, Buffalo County from July 8, 1998 to May 21 1999, at a feeder in Hastings, Adams County from May to June 2009 and in August 2009, in Kearney, Buffalo County, Hastings, Adams County, and Grand Island, Hall County in June 2010. A single bird was seen at a feeder in Hastings, Adams County on June 8, 2011 (Sharpe et al., 2001; Jorgensen, 2012)
County Records: [A, B, D, L, P]

Mourning Dove, *Zenaida macroura*
Abundant permanent resident and regular breeder throughout the region. less common in winter.
County Records: [A, B, C, D, F, G, HL, HM, K, L, M, P]

Mourning Dove, adult

Inca Dove, *Columbina inca*
Very rare fall and winter visitor. Observed in Kearney, Buffalo County from October 28, 1987 to March 7, 1988 (Paine, 1988; NBR 56:3).
County Records: [B]

Common Ground Dove, *Columbina passerina*
Very rare winter visitor. Observed at Hanson WPA, Clay County on November 26, 2004 (Jorgensen, 2012).
County Records: [C]

Order Psittaciformes

Family Psittacidae: Parrots

[Monk Parakeet, *Myopsitta monachus*]
Hypothetical; an escaped captive was seen in Kearney, Buffalo County, May 6, 1975 (Bliese, 1975; NBR 43:42).
County Records: [B]

[**Budgerigar**, *Melopsittacus undulatus*]
Hypothetical, an escaped captive was seen in Kearney, Buffalo County, August 8, 1991 (Sharpe et al., 2001; NBR 60: 32).

Order Cuculiformes

Family Cuculidae: Cuckoos

Yellow-billed Cuckoo, *Coccyzus americanus*
Uncommon summer resident and regular breeder throughout the region. To be expected from late April to early November. Found in brushy areas with scattered trees or woodlands with thorny bushes near water. Tier 2 At-Risk species.
County Records: [A, B, C, D, F, G, HL, HM, K, L, M, P]

Black-billed Cuckoo, *Coccyzus erythropthalmus*
Uncommon summer resident and regular breeder, to be expected from early May to early October. Occurs in brushy areas with scattered trees; may prefer more wooded habitats than does the Yellow-billed Cuckoo. Tier 2 At-Risk species.
County Records: [A, B, HL, L]

Groove-billed Ani, *Crotophaga sulcirostris*
Very rare fall visitor. Seen October 15 November 1, 1975 in Hall County (Stoppkotte, 1975).
County Records: [HL]

Order Strigiformes

Family Tytonidae: Barn Owls

Barn Owl, *Tyto alba*
Uncommon year-round resident and regular breeder, less frequently seen in winter. Numbers may be declining in area due to the loss of abandoned farm buildings and holes in banks used for nest sites. Tier 2 At-Risk species.
County Records: [A, B, F, L]

Yellow-billed Cuckoo, adult

Barn Owl, adult

Family Strigidae: Typical Owls

Flammulated Owl, *Otus flammeolus*
Very rare visitor, one specimen collected near Kearney, Buffalo
County in 1891 (Sharpe et al., 2001). ABC Yellow-List species.
County Records: [B]

Eastern Screech-Owl, *Megascops asio*
Uncommon year-round resident and regular breeder, local in west.
Species benefited from planting of windbreaks, but may decline as
windbreaks are removed. *M. a. maxwelliae* occurs in region, with
three color morphs: gray (common), red (uncommon) and intermediate brown (rare) (Sharpe et al., 2001).
County Records: [A, B, C, D, F, G, HL, HM, K, L, M, P]

Great Horned Owl, *Bubo virginianus*
Common year-round resident and regular breeder throughout the
region; may disperse in winter if food resources are scarce. Uses
large trees for breeding, often usurping crow and hawk nests, but
also nests under bridges and on cliffs. *B. v. virginianus* is most common in region, but *B. v. occidentalis*, *B. v. wapacuthu* and *B. v. lapo-
phonus* are also possible (Swenk, 1937; Sharpe et al., 2001).
County Records: [A, B, C, D, F, G, HL, HM, K, L, M, P]

Snowy Owl, *Bubo scandiaca*
Very rare winter visitor, to be expected from mid-November to mid-
March. The winter of 2011–2012 was a record-breaking invasion year
with reports of approximately 200 individuals seen across Nebraska
(J. G. Jorgensen, pers. comm.); 1917–1918 was also a major invasion
year (Sharpe et al., 2001).
County Records: [B, HL, HM, K, L, P]

Burrowing Owl, *Athene cunicularia*
Uncommon spring and fall migrant, to be expected from late
March to late October; rare local breeder. Typically found near prai-
rie-dog or other small mammal burrows in open grassland, prairies and pastures. Declining due to loss of habitat (Bruner 1904),
and the widespread poisoning that often severely affects non-tar-
get species and others that use the burrows commensally. Tier 1 At-
Risk species.
County Records: [A, B, F, HL, K, L, M, P]

Barred Owl, *Strix varia*
Very rare summer visitor. Observed on October 3, 1962 in Dawson County, February 4, 1974 and March 10, 2008 in Adams County, on June 2, 1991 in Hall County and reported in Lincoln County (Tout, 1947; Sharpe et al., 2001; Jorgensen, 2012; NBR 42: 74). Tier 2 At-Risk species.
County Records: [A, D, HL, L]

Great Gray Owl, *Strix nebulosa*
Very rare winter visitor. Reported in Frontier County in 1896 and near Brady, Lincoln County in 1950 (Sharpe et al., 2001; NBR 18:83).
County Records: [F, L]

Burrowing Owl, adult

Long-eared Owl, *Asio otus*
Uncommon year-round resident, with some spring and fall migrants and rare, local breeder, migrants to be expected from early March and late April and late October and early November. Tier 2 At-Risk species.
County Records: [A, B, HL, L, P]

Short-eared Owl, *Asio flammeus*
Uncommon year-round resident with some spring and fall migrants, winter visitor and rare, local breeder. May be more common in region than has been reported. Tier 2 At-Risk species and ABC Yellow-List species.
County Records: [A, B, C, G, HL, K. L, P]

Northern Saw-whet Owl, *Aegolius acadicus*
Uncommon fall and winter visitor. A specimen was collected near Inland, Clay County on November 10, 1917 and another was collected at Hastings, Adams County during the winter of 1922–1923 (Swenk, 1925; Jorgensen, 2012). From 2004–2006, 40 owls, both adults and young, were banded in the vicinity of the Crane Trust in Hall County.
County Records: [A, C, L]

Boreal Owl, *Aegolius funearius*
Very rare fall visitor. One individual was collected at Inland, Clay County on October 5, 1916 (Jorgensen, 2012).
County Records: [A, B, C]

Order Caprimulgiformes

Family Caprimulgidae: Goatsuckers

Common Nighthawk, *Chordeiles minor*
Common spring and fall migrant and regular breeder throughout the region, to be expected from mid-April to late October. Migrants tend to arrive earlier in the east than in the west. *C. m. minor* and *C. m. sennetti* may occur in region (Haecker et al., 1945; AOU, 1998; Sharpe et al., 2001).
County Records: [A, B, C, D, F, G, HL, HM, K, L, M, P]

Common Poorwill, *Phalaenoptilus nuttallii*
Rare spring and fall migrant and rare summer visitor, migrants to be expected from late April to early October. Early reports include May 2, 1901 in Kearney, Buffalo County, June 6, 1913 in Lincoln County, fall 1914 at Funk, Phelps County and September 20, 1930 and April 26, 1951 in Adams County. Tier 2 At-Risk species.
County Records: [A, B, F, HL, L, P]

Chuck-will's-widow, *Caprimulgus carolinensis*
Very rare spring migrant, to be expected from early May to late June. Found along the Platte River between Kearney, Buffalo County and Minden, Kearney County in June 19–29, 1983 and in 1997 (Kimball, 1984; Sharpe et al., 2001; NBR 65: 108). Tier 2 At-Risk species.
County Records: [B, K]

Whip-poor-will. *Caprimulgus vociferus*
Very rare spring migrant, to be expected from mid-April to late May. One individual was collected near Giltner, Hamilton County in May 1896, Swenk (1925) reported hearing birds near Harvard, Clay County and Turner (1934) reported the species near Holstein, Adams County on May 18, 1934 (Jorgensen, 2012). Tier 2 At-Risk species.
County Records: [A, C, D, HM]

Order Apodiformes

Family Apodidae: Swifts

Chimney Swift, *Chaetura pelagica*
Very common spring and fall migrant, summer resident and regular breeder, migrants to be expected from mid-April to mid-October. Typically occurs near human structures and may be declining due to the loss of these nesting and roosting sites.
County Records: [A, C, D, F, G, HL, HM, K, L, M, P]

Family Trochilidae: Hummingbirds

Ruby-throated Hummingbird, *Archilochus colubris*
Occasional spring and uncommon fall migrant and rare summer visitor, migrants to be expected from mid-April to late May, and

early August to mid-October. Distribution largely restricted to east of the 96th longitude meridian (Sharpe et al., 2001), but backyard bird feeders are increasing the numbers seen across the region. Swenk (1925) considered ruby-throats rare at Inland, Clay County, but they have been regularly seen in Hastings, Adams County since the 1930's (Jorgensen, 2012). Tier 2 At-Risk species.
County Records: [A, B, C, HL, L]

Costa's Hummingbird, *Calypte costae*
Very rare late summer and early fall migrant. ABC Yellow-List species.
County Records: [D]

Calliope Hummingbird, *Stellula calliope*
Very rare spring and late summer-early fall migrant, to be expected from April and late July to August, often feeding on *Penstemon* flowers (Viehmeyer, 1961; Shickley, 1965; NBR 80: 55). ABC Yellow-List species.
County Records: [D, L]

Broad-tailed Hummingbird, *Selasphorus platycercus*
Rare late summer and early fall migrant, to be expected from late July to early September. Individuals seen in Hastings, Adams County on August 22–30, 1987 and August 21, 1965 (Grenon, 1990; Sharpe et al., 2001; Jorgensen, 2012; NBR 33: 39). There are early reports, August 1903, July 1914 and August 1921 from Kearney, Buffalo County, and August 1959 and September 1960 from Lincoln County (Sharpe et al., 2001).
County Records: [A, B, L, P]

Rufous Hummingbird, *Selasphorus rufus*
Very rare late summer and early fall migrant, to be expected from mid-July to mid-September. An adult male was seen in Hastings, Adams County on September 13–14, 1987 and one seen in Lincoln County on August 15, 1992 (Sharpe et al., 2001; Jorgensen, 2012; NBR 56:14).
County Records: [A, L]

Order Coraciiformes

Family Alcedinidae: Kingfishers

Belted Kingfisher, *Ceryle alcyon*
Common year-round resident and uncommon local breeder throughout the region. with some spring and fall migrants. Migrants to be
expected from mid-April to early October.
County Records: [A, B, C, D, F, G, HL, HM, K, L, M, P]

Order Piciformes

Family Picidae: Woodpeckers

Lewis's Woodpecker, *Melanerpes lewis*
Very rare spring, summer and winter visitor. One individual was
seen on May 30, 1943 and another seen on January 20, 1954 in Hast-
ings, Adams County (NBR 11:41; NBR 22:59), one seen near Kearney,
Buffalo County on May 16 1900 (Swenk, 1925) and one in Lincoln
County in July (Tout, 1947). Tier 1 At-Risk species and ABC Red-List
species.
County Records: [A, B, L]

Red-headed Woodpecker, *Melanerpes erythrocephalus*
Very common spring and fall migrant, summer resident and very
rare winter visitor, migrants to be expected from mid-April to late
November. Declining in other parts of its range due to nesting habitat loss as wooded areas converted to agriculture. ABC Yellow-List
species.
County Records: [A, B, C, D, F, G, HL, HM, K, L, M, P]

Red-bellied Woodpecker, *Melanerpes carolinus*
Common year-round resident and regular breeder throughout the
region.. Range expanded westward in the 20th century. Occurred
only in Otoe and Cass counties in 1920's (Ducey 1988; Sharpe et al.
2001); at Hastings, Adams County in 1930's; at North Platte, Lincoln County in 1950; and at Sutherland, Lincoln County in 1956–
1957 (Sharpe et al. 2001).
County Records: [A, B, C, D, F, G, HL, HM, K, L, M, P]

Williamson's Sapsucker, *Sphyrapicus thyroideus*
Very rare spring visitor. A male was seen in Heartwell Park, Hastings, Adams County on March 24, 1939 (Jones, 1939) and another in Grand Island, Hall County on May 5, 1959 (NBR 27: 51). ABC Yellow-List species.
County Records: [A, HL]

Yellow-bellied Sapsucker, *Sphyrapicus varius*
Occasional spring and fall migrant and winter visitor, to be expected from late September to early May.
County Records: [A, B, F, HL, HM, K, L]

Downy Woodpecker, *Picoides pubescens*
Abundant year-round resident and regular breeder throughout the region; may move to larger wooded areas during the winter, but wanders less than Hairy Woodpecker. *P. p. medianus* occurs in the region (Sharpe et al., 2001).
County Records: [A, B, C, D, F, G, HL, HM, K, L, M, P]

Hairy Woodpecker, *Picoides villosus*
Common year-round resident and regular breeder. Requires larger wooded areas than Downy Woodpecker. Competition with Northern Flickers may discourage birds from using an area. Considered common at Inland, Clay County in early 1900's (Swenk, 1925; Jorgensen, 2012), but declining across region since 1950s due to nesting habitat loss. *P. v. villosus* occurs in region (Sharpe et al., 2001).
County Records: [A, B, C, D, F, G, HL, HM, K, L, M, P]

Northern Flicker, *Colaptes auratus*
Abundant year-round resident and regular breeder throughout the region. Both red-shafted (*cafer*) and yellow-shafted (*auratus*) races occur in the region, along with hybrids; Short (1961) believed that most summer residents are hybrids.
County Records: [A, B, C, D, F, G, HL, HM, K, L, M, P]

2. Passerine Families

Order Passeriformes

Family Tyrannidae: American Flycatchers

Olive-sided Flycatcher, *Contopus cooperi*
Occasional spring and fall migrant and rare summer resident, migrants to be expected from late April to mid-June and early August to early October. ABC Yellow-List species.
County Records: [A, B, G, HL, L, P]

Western Wood-Pewee, *Contopus sordidulus*
Rare spring migrant and rare summer resident, migrants to be expected from early May to mid-June. Observed in Adams County on May 28, 1973 (NBR 41: 55) and May 18, 1985 (NBR 53: 58), in Lincoln County on June 18, 2010 and on July 6, 2011. Similarity with Eastern Wood-Pewee makes the species' range limits difficult to establish.
County Records: [A, L]

Eastern Wood-Pewee, *Contopus virens*
Occasional spring and fall migrant and rare summer resident and regular breeder, migrants to be expected from late April through early October. Number of birds declined in the 1950's due to the loss of American elms (*Ulmus americana*) in the region (Sharpe et al., 2001). Similarity with Western Wood-Pewee makes the species' range limits difficult to establish, but it seems to be expanding westward through the region.
County Records: [A, B, C, F, HL, HM, K, L, M, P]

Yellow-bellied Flycatcher, *Empidonax flaviventris*
Uncommon spring migrant and rare summer resident, migrants to be expected from early May to early June and early September. All *Empidonax* flycatchers in the region should be observed closely to help establish the distributions of the several species in the region.
County Records: [A, HM]

[Acadian Flycatcher, *Empidonax virescens*]
Hypothetical spring migrant and rare summer resident, migrants to be expected from early to mid-May. Brookings reported seeing

Acadian Flycatchers, but later examination suggested they were "Traill's" (Willow) Flycatchers (Swenk, 1925; Sharpe et al., 2001). Tier 2 At-Risk species.
County Records: [A, C, HL, L]

Alder Flycatcher, *Empidonax alnorum*
Occasional spring migrant, migrants to be expected from early to mid-May and late July to early September.
County Records: [A, HL, K]

Willow Flycatcher, *Empidonax traillii*
Abundant spring migrant, abundant summer resident and regular breeder, migrants to be expected from early May to September. Literature prior to 1973 lumped this species with the Alder Flycatcher as the "Trail's" Flycatcher. ABC Yellow-List species.
County Records: [A, B, C, D, G, HL, HM, K, L, M, P]

Least Flycatcher, *Empidonax minimus*
Common spring and fall migrant and common summer resident, migrants to be expected from late April to mid-May, and late September to early October. This is the earliest of the *Empidonax* flycatchers to arrive in spring.
County Records: [A, B, C, D, F, G, HL, HM, K, L, M, P]

Hammond's Flycatcher, *Empidonax hammondii*
Very rare fall visitor. One bird was observed at the DLD State Wayside Area 2, Adams County in September 1995 (Gubanyi, 1996b).
County Records: [A]

Eastern Phoebe, *Sayornis phoebe*
Common spring migrant and common summer resident and regular breeder throughout the region, migrants to be expected from mid-March to mid-November. This is the earliest flycatcher to arrive in spring. Range expanding westward due to the increasing number of artificial structures being used as nesting substrate.
County Records: [A, B, C, D, F, G, HL, HM, K, L, M, P]

Say's Phoebe, *Sayornis saya*
Uncommon spring migrant and uncommon summer resident and regular breeder throughout the region, migrants to be expected from late March to late May, and early to mid-September. More common in western counties, Bruner et al. (1904) reported that the

species' eastward spread was probably due to the increasing number of artificial structures the birds use as nesting substrate.
County Records: [A, B, D, F, G, HL, HM, K, L, M, P]

Vermillion Flycatcher, *Pyrocephalus rubinus*

Very rare spring and fall visitor. Seen in North Platte, Lincoln County from October to December 1954, on November 4, 1960 and on March 27, 1976 (Rapp, 1958; Bray et al., 1986; Sharpe et al., 2001).
County Records: [L]

Great Crested Flycatcher, *Myiarchus crinitus*

Very common spring migrant and summer resident throughout the region, migrants to be expected from mid-April to early August. Bruner et al. (1904) reported species to be restricted to southeast Nebraska along the Missouri River and in Lancaster County; within 20 years Swenk and Dawson (1921) reported species observed as far west as Adams County.
County Records: [A, B, C, D, F, G, HL, HM, K, L, M, P]

Cassin's Kingbird, *Tyrannis vociferans*

Very rare spring visitor. Observed in North Platte, Lincoln County on May 6, 1961 and May 21 1961 (NBR 29: 48, 36:56). Tier 2 At-Risk species
County Records: [L]

Western Kingbird, *Tyrannus verticalis*

Abundant spring and fall migrant and regular breeder throughout the region,, migrants to be expected from early April to early October. Generally found in habitats that are more open and drier than Eastern Kingbird's.
County Records: [A, B, C, D, F, G, HL, HM, K, L, M, P]

Eastern Kingbird, *Tyrannus tyrannus*

Abundant spring and fall migrant and regular breeder, migrants to be expected from early March to early October.
County Records: [A, B, C, D, F, G, HL, HM, K, L, M, P]

Scissor-tailed Flycatcher, *Tyrannus forficatus*

Very rare spring migrant and summer visitor, migrants to be expected from mid-April to late May. One was seen in Lincoln County on July 29, 2009, and a nest was found in Kearney County on June

30, 2010. It may have nested near Hastings, Adams County in 1939 and 1943, and near Harvard, Clay County in 1964 (Jorgensen, 2012). Tier 2 At-Risk species.
County Records: [A, B, C, HM, K, L]

Family Laniidae: Shrikes

Loggerhead Shrike, *Lanius ludovicianus*
Uncommon spring and fall migrant throughout the region. A year-round resident and regular breeder, migrants to be expected from mid-February to late October. *L. l. excubitorides* or intergrades with *L. l. migrans* occur in region (Sharpe et al., 2001). Range wide, the species is declining and its range contracting due to habitat loss and pesticide use (Robins and Easterla, 1992). Tier 2 At-Risk species
County Records: [A, B, C, D, F, G, HL, HM, K, L, M, P]

Northern Shrike, *Lanius excubitor*
Uncommon spring and fall migrant and winter visitor throughout the region,, to be expected from mid-September to mid-May. *L. e. invictus* or intergrades with *L. e. borealis* occur in region (Sharpe et al., 2001).
County Records: [A, B, C, D, F, G, HL, HM, K, L, M, P]

Family Vireonidae: Vireos

Bell's Vireo, *Vireo bellii*
Common spring and fall migrant, summer resident and regular breeder throughout the region, migrants to be expected late April to late May, and early August to late September. Tier 1 At-Risk species and ABC Red-List species.
County Records: [A, B, C, D, F, G, HL, HM, K, L, M, P]

Yellow-throated Vireo, *Vireo flavifrons*
Rare spring and fall migrant and rare summer resident and breeder, migrants to be expected mid-April to late May, and late September to late October. Tier 2 At-Risk species.
County Records: [A, C, HL, HM, L]

Cassin's Vireo. *Vireo cassinii*
Very rare fall migrant, to be expected mid-August to mid-September. Very few records exist for this recently (1998) recognized species.
County Records: [D, F]

Northern Shrike, adult

Blue-headed Vireo, *Vireo solitarius*
Uncommon spring and fall migrant; migrants to be expected late April to mid-June, and late August to mid-October. The Solitary Vireo complex was recently separated into Blue-headed, Plumbeous (*V. plumbeus*), and Cassin's (AOU, 1998); additional observations will be necessary to evaluate the relative occurrence of the three species in the region. There are no regional records for the Plumbeous, but it is an uncommon breeder in the Black Hills, and migrants might be expected.
County Records: [A, B, HL, M]

Warbling Vireo, *Vireo gilvus*
Abundant spring and fall migrant, summer resident and regular breeder throughout the region,, migrants to be expected late April to late September. *V. g. gilvus* occurs in the region (Haecker et al., 1945; AOU, 1998).
County Records: [A, B, C, D, F, G, HL, HM, K, L, P]

Philadelphia Vireo, *Vireo philadelphicus*
Uncommon spring and fall migrant and summer resident, migrants to be expected late April to early June and late August to early October.
County Records: [A, C, HL, L]

Red-eyed Vireo, *Vireo olivaceus*
Rare spring and fall migrant, rare summer resident and regular breeder throughout the region,, migrants to be expected late April to late May, and late July to mid-October. Tout (1902) considered the species common near Inland, Clay County in the early 1900's.
County Records: [A, B, C, D, F, HL, HM, K, L, M, P]

Red-eyed Vireo, adult at nest

White-eyed Vireo, *Vireo griseus*
Very rare spring and fall migrant, to be expected mid-April to late May, and mid-September. Brooking collected one individual near Inland, Clay County on May 20, 1917. Single birds seen in Adams County on May 19, 1928 (LOI 33:6) and May 31, 1974 (NBR 42:79). County Records: [A, C]

Family Corvidae: Jays, Magpies, and Crows

Pinyon Jay, *Gymnorhinus cyanocephalus*
Rare fall visitor, to be expected in mid-to late October. The species current range is much the same as 100 years ago (Bruner et al. 1904). *G. c. cyanocephalus* occurs in region (Sharpe et al., 2001). Tier 2 At-Risk species and ABC Yellow-List species. County Records: [A, B, C, D, L]

Steller's Jay, *Cyanocitta stelleri*
Very rare fall and winter visitor, to be expected mid-August to late April. One seen near Kearney, Buffalo County on September 13, 1914 (**Swenk? 1925**) and one seen near North Platte, Lincoln County on October 15, 1936. Also seen near North Platte, Lincoln County during the winters of 1947 and 1962–1963 (Sharpe et al., 2001). County Records: [B, L]

Blue Jay, *Cyanocitta cristata*
Abundant year-round resident and regular breeder throughout the region. Distribution has not changed since the report of Bruner et al. (1904). *C. c. cyanotephra* occurs in region (Sharpe et al., 2001). The 2002–2003 West Nile Virus had little if any lasting effects on this species. County Records: [A, B, C, D, F, G, HL, HM, K, L, M, P]

Western Scrub-Jay, *Aphelocoma californica*
Very rare, Bruner (1896) reports the species as a "common transient visitor" in Lincoln County, but may have been referring to Pinyon Jays (NBR 48:89). County Records: [L]

Clark's Nutcracker, *Nucifraga columbiana*
Very rare fall and winter visitor. Sightings reported in 1899 and 1955 (Bruner, 1904; Sharpe et al., 2001). County Records: [B, L, M]

Blue Jay, adult

Black-billed Magpie, *Pica hudsonia*
Uncommon spring and fall migrant, year-round resident and regular breeder, mainly westwardly. Migrants to be expected early March to late April, and early October to late November. Population numbers were reduced by coyote poisoning campaigns in the late 1800's–early 1900's and more recently (2002–2003) by West Nile Virus, from which the central Platte population has yet to recover. County Records: [A, B, C, D, F, G, HL, HM, K, L, M, P]

American Crow, adult

American Crow, *Corvus brachyrhynchos*
Very common year-round resident and regular breeder through-out the region. In the early 1900's, species was rare west of the 98th meridian, but range moving westward (Bruner, 1904). Popula-tion numbers reduced by West Nile Virus in 2002–2003, but now recovering.
County Records: [A, B, C, D, F, G, HL, HM, K, L, M, P]

Chihuahuan Raven, *Corvus cryptoleucus*
Very rare spring migrant and very rare local breeder. A nest was found between Wilcox and Axtell, Kearney County on June 11, 1944 (Brooking, 1944) and another was seen south of Axtell, Kear-ney County on April 13, 1947 (Brown, 1947). Birds were seen near Holstein, Adams County on April 25, 1927 (Sharpe et al., 2001). Bi-

son carcasses provided a dependable food resource for ravens; both species largely disappeared when the great bison herds disappeared (Bent 1946; Andrews and Righter, 1992; Sharpe et al., 2001). Known as the White-necked Raven prior to 2000.
County Records: [A, HL, K, L]

Common Raven, *Corvus corax*
Very rare spring and fall visitor. Very few sightings after 1900; two reports from Adams County on May 21, 1952 and March 21, 1962 (NBR 30:65) and one from Buffalo County in 1960 (Sharpe et al., 2001). Like the previous species, the Common Raven was extirpated from Nebraska in the late 1800's, but there have been a few scattered recent sightings in the state.
County Records: [A, B]

Top to bottom, left to right: Horned Lark, Rock Wren, Chestnut-collared Longspur (male), and Brown-headed Cowbird (male)

Family Alaudidae: Larks

Horned Lark, *Eremophila alpestris*
Abundant spring and fall migrant, common year-round resident and regular breeder throughout the region. *E. a. anthymia* breeds in region (AOU, 1998), but other races as likely as migrants.
County Records: [A, B, C, D, F, G, HL, HM, K, L, M, P]

Family Hirundinidae: Swallows and Martins

Purple Martin, *Progne subis*
Common spring and fall migrant and uncommon summer resident throughout the region. A regular breeder, to be expected late March to early October. Entirely dependent on artificial nesting structures (martin houses and gourds), and competes with European Starlings and House Sparrows for access to these structures.
County Records: [A, B, C, D, F, G, HL, HM, K, L, M, P]

Tree Swallow, *Tachycineta bicolor*
Common spring and fall migrant, summer resident and regular breeder throughout the region, to be expected mid-March to late October. Nebraska is near the historic western edge of breeding range, but it is expanding westward with nest boxes and aging woodlots providing nesting habitat.
County Records: [A, B, C, D, F, G, HL, HM, K, L, M, P]

Violet-green Swallow, *Tachycineta thalassina*
Very rare spring and fall migrant and very rare summer visitor, to be expected late April to mid-May and early to late August. Tier 2 At-Risk species.
County Records: [A, L]

Northern Rough-winged Swallow, *Stelgidopteryx serripennis*
Abundant spring and fall migrant, summer resident and regular breeder throughout the region, to be expected early April to mid-October.
County Records: [A, B, C, D, F, G, HL, HM, K, L, M, P]

Bank Swallow, *Riparia riparia*
Abundant spring and fall migrant, abundant summer resident and regular breeder throughout the region, to be expected mid-April to early October.
County Records: [A, B, C, D, F, G, HL, HM, K, L, M, P]

Tree Swallow, adult male at nest

Cliff Swallow, *Petrochelidon pyrrhonota*
    Abundant spring and fall migrant, abundant summer resident and
    regular breeder throughout the region, to be expected mid-April to
    late September. Historically placed nests on sides of cliffs but nest-
    ing habitat now consists of artificial structures located near water
    (bridges, road culverts or irrigation structures). Highly social spe-

cies, forming nesting colonies of up to thousands of individuals (Brown and Brown, 1996).
County Records: [A, B, C, D, F, G, HL, HM, K, L, M, P]

Barn Swallow, *Hirundo rustica*
Abundant spring and fall migrant, abundant summer resident and regular breeder throughout the region, to be expected early April to early November. Historically placed nests on sides of cliffs but nesting habitat now consists of artificial structures (barns, farm buildings).
County Records: [A, B, C, D, F, G, HL, HM, K, L, M, P]

Family Paridae: Chickadees and Titmice

Black-capped Chickadee, *Poecile atricapillus*
Abundant spring and fall migrant, year-round resident and regular breeder throughout the region Distribution in the region unchanged since the early 1900's (Bruner, 1904). Population reduced by West Nile Virus in 2002-2003, but now recovering. *P. a. septentrionalis* occurs in region (Bruner, 1904).
County Records: [A, B, C, D, F, G, HL, HM, K, L, M, P]

Black-capped Chickadee, adult

Mountain Chickadee, *Poecile gambeli*
Very rare fall and winter visitor. Seen on November 6, 1968, April 6, 1969 and November 16, 1976 in Lincoln County, and during January–March 1969 in Dawson County.
County Records: [D, L]

Tufted Titmouse, *Baeolophus bicolor*
Very rare spring, fall and winter visitor, to be expected mid-February to mid-April. Tier 2 At-Risk species.
County Records: [A, C, HL, HM, L]

Family Sittidae: Nuthatches

Red-breasted Nuthatch, *Sitta canadensis*
Uncommon spring and fall migrant and common winter resident throughout the region, to be expected early August to late May.
County Records: [A, B, C, D, F, G, HL, HM, K, L, M, P]

White-breasted Nuthatch, *Sitta carolinensis*
Common spring and fall migrant, year-round resident and regular breeder throughout the region. Historically only occurred in the eastern third of the state (Bruner et al., 1904; Haecker et al., 1945; Rapp et al., 1958), but range expanding westward in recent decades. *S. c. cookei* occurs in the region (AOU, 1998). Tier 2 At-Risk species.
County Records: [A, B, C, D, F, G, HL, HM, K, L, M, P]

Pygmy Nuthatch (*Sitta pygmaea*)
Very rare winter visitor. Observed near Minden, Kearney County from January 20–April 20, 1983 (Jorgensen, 2012; NBR 51:91), and in Lincoln County in September 1976 (Sharpe et al., 2001).
County Records: [K, L]

Family Certhiidae: Creepers

Brown Creeper, *Certhia americana*
Uncommon spring and fall migrant, uncommon winter resident and rare, local breeder throughout the region, to be expected late September to late May. *C. a. americana* and *C. a. montana* are both possible in the region (AOU, 1998). Tier 2 At-Risk species.
County Records: [A, B, C, D, G, HL, HM, K, L, M, P]

White-breasted
Nuthatch, adult

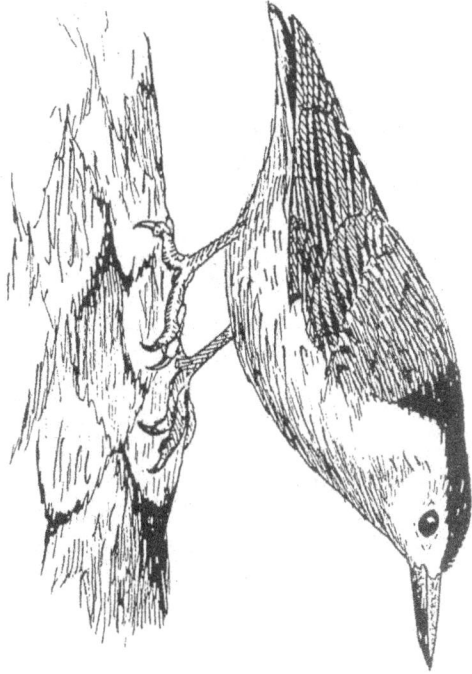

Brown Creeper, adult

Family Troglodytidae: Wrens

Rock Wren, *Salpinctes obsoletus*
Rare spring and summer visitor, to be expected early April to mid-October. More commonly seen in the west, but range has been expanding eastward recently (Sharpe et al., 2001).
County Records: [A, B, C, D, F, HL, L]

Carolina Wren, *Thryothorus ludovicianus*
Rare year-round resident and local breeder More commonly seen in eastern Nebraska, but range has been expanding westward. Tier 2 At-Risk species.
County Records: [A, B, C, D, G, HL, L]

Bewick's Wren, *Thryomanes bewickii*
Rare spring and fall migrant, rare summer resident and local breeder and very rare winter visitor, migrants to be expected early April to late May, and early September to late October. *T. b. bewicki* occurs in region, but *T. b. pulichi* also possible (Sharpe et al., 2001).
County Records: [A, B, C, HL, L]

House Wren, *Troglodytes aedon*
Abundant spring and fall migrant, abundant summer resident and very rare winter visitor throughout the region, to be expected early April to late October.
County Records: [A, B, C, D, F, G, HL, HM, K, L, M, P]

Winter Wren, *Troglodytes troglodytes*
Rare spring and fall migrant and rare summer and winter visitor, to be expected mid-to late May, and mid-September to late January. *T. t. hiemalis* occurs in the region. In light of the Winter Wren–Pacific Wren taxonomic split and difficulties of identification, careful scrutiny of "Winter Wrens" seen in the region is warranted.
County Records: [A, B, D, HL, L, P]

Sedge Wren, *Cistothorus platensis*
Common spring and fall migrant, rare summer resident and regular breeder, migrants to be expected mid-April to early June, and mid-July to late October. Tier 2 At-Risk species.
County Records: [A, B, C, G, HL, HM, K, L, M, P]

Marsh Wren, *Cistothorus palustris*
Common spring migrant and uncommon summer resident and

regular breeder, migrants to be expected mid-April to mid-May, and early September to late November.
County Records: [A, B, C, D, F, G, HL, HM, K, L, M, P]

Family Polioptilidae: Gnatcatchers

Blue-gray Gnatcatcher, *Polioptila caerulea*
Occasional spring and fall migrant, rare summer resident and local breeder, migrants to be expected early April to mid-May, and late June to late September. May be increasing in the region as range expands westward (Jorgensen, 2012); most, if not all, of the Nebraska's southern tier of counties have been colonized, as have some of the westernmost counties (W. Mollhoff, pers. comm.). Tier 2 At-Risk species.
County Records: [A, C, HL, L]

Family Regulidae: Kinglets

Golden-crowned Kinglet, *Regulus satrapa*
Uncommon spring and fall migrant and uncommon winter visitor throughout the region, migrants to be expected early March to mid-May, and mid-September to late November.
County Records: [A, B, C, D, F, G, HL, HM, K, L, P]

Golden-crowned Kinglet, adult male

Ruby-crowned Kinglet, *Regulus calendula*
Common spring and fall migrant and rare winter visitor throughout the region, migrants to be expected early March to late May,
and mid-August to early December.
County Records: [A, B, C, D, F, G, HL, HM, K, L, M, P]

Family Turdidae: Thrushes and Allies

Eastern Bluebird, *Sialia sialis*
Abundant spring and fall migrant, common year-round resident
and regular breeder throughout the region, to be expected mid-
March to mid-November. Nesting occurs in nest boxes, woodpecker holes and tree snags. Range in Nebraska largely unchanged
over the past 100 years (Bruner et al., 1904).
County Records: [A, B, C, D, F, G, HL, HM, K, L, M, P]

Mountain Bluebird, *Sialia currucoides*
Uncommon spring migrant, rare winter visitor and very rare local
breeder, migrants to be expected mid-February to mid-May and
late June to mid-November. Tier 2 At-Risk species.
County Records: [D, F, HL, K, L, M]

Townsend's Solitaire, *Myadestes townsendi*
Uncommon spring and fall migrant and occasional winter visitor,
to be expected late November to early May. Tier 2 At-Risk species.
County Records: [A, B, D, F, HL, HM, K, L, P]

Veery, *Catharus fuscescens*
Occasional spring and fall migrant, to be expected late April to
early June and early to late September. *C. f. fuscescens* occurs in region (Swenk, 1925).
County Records: [C, HL, K, L]

Gray-cheeked Thrush, *Catharus minimus*
Occasional spring and fall migrant, to be expected late April to
early June and late August to mid-October.
County Records: [A, D, HL, L]

Swainson's Thrush, *Catharus ustulatus*
Common spring and fall migrant throughout the region, to be expected late April to mid-June and late August to late October. *C. u.
swansoni* occurs in region (Sharpe et al., 2001).
County Records: [A, B, C, D, F, HL, HM, K, L, M, P]

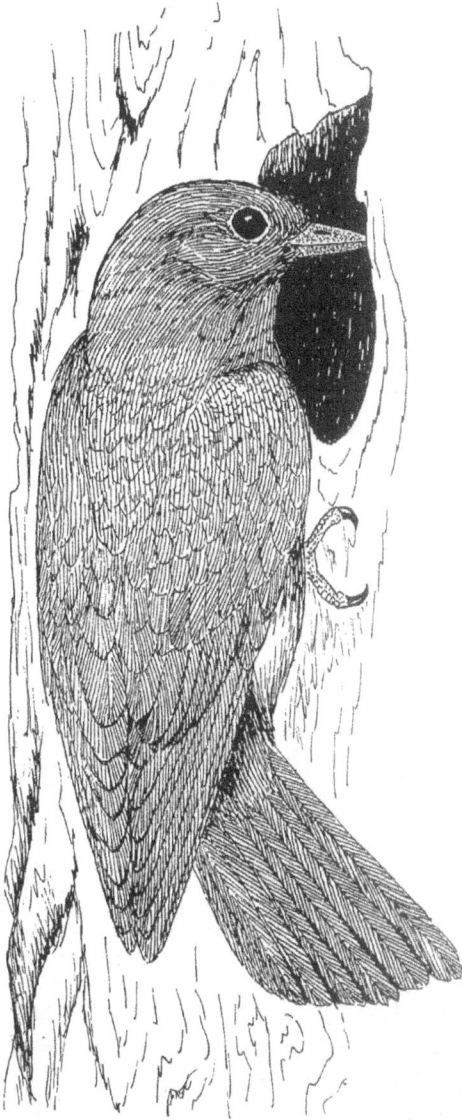

Eastern Bluebird, adult female at nest

Hermit Thrush, *Catharus guttatus*
  Rare spring and fall migrant, to be expected late March to mid-May
  and mid-September to early November. *C. g. faxoni* occurs in region (Sharpe et al., 2001).
  County Records: [A, HL, L]

Wood Thrush, *Hylocichla mustelina*
Uncommon spring and fall migrant, rare summer resident and lo-
cal breeder, migrants to be expected late April to late May and mid-
September to mid-October. ABC Yellow-List species.
County Records: [A, B, F, HL, HM, L, M]

American Robin, *Turdus migratorius*
Abundant spring and fall migrant, abundant year-round resident
and regular breeder throughout the region, migrants to be ex-
pected early February to early April and late July to mid-December.
T. m. migratorius occurs in region (Sharpe et al., 2001).
County Records: [A, B, C, D, F, G, HL, HM, K, L, M, P]

Varied Thrush, *Ixoeus naevius*
Very rare spring, fall and winter visitor. One seen in North Platte,
Lincoln County on December 15, 1936 (Tout, 1947) and in Kearney,
Buffalo County on May 17, 2001. ABC Yellow-List species.
County Records: [B, L]

Family Mimidae: Mockingbirds, Thrashers, and Catbirds

Gray Catbird, *Dumetella carolinensis*
Common spring and fall migrant, uncommon summer resident,
regular breeder and very rare winter visitor, migrants to be expected
mid-April to mid-May and mid-July to early November.
County Records: [A, B, C, D, F, G, HL, HM, K, L, P]

Northern Mockingbird, *Mimus polyglottos*
Occasional spring and fall migrant, rare year-round resident
throughout the region. A regular breeder and occasional winter visitor, to be expected late April to mid-September. *M. p. leucopterus*,
M. p. polyglottos and intergrades are possible in region (Sharpe et
al., 2001).
County Records: [A, B, C, D, F, G, HL, HM, K, L, P]

Sage Thrasher, *Oreoscoptes montanus*
Very rare summer visitor. Observed in Lincoln County on July 4,
1942 (Tout, 1947; Sharpe et al., 2001).
County Records: [L]

Brown Thrasher, *Toxostoma rufum*
Common spring and fall migrant, abundant year-round resident and regular breeder throughout the region, migrants to be expected early April to mid-May and late July to late October. *T. r. longicauda* and *T. r. rufum* are possible in region (Sharpe et al., 2001).
County Records: [A, B, C, D, F, G, HL, HM, K, L, M, P]

Curve-billed Thrasher, *Toxostoma curvirostra*
Very rare spring visitor. Observed in North Platte, Lincoln County from April 19 to May 3, 1936 (Weakley, 1936; Sharpe et al., 2001).
County Records: [L]

Family Sturnidae: Starlings

European Starling, *Sturnus vulgaris*
Abundant year-round resident and regular breeder throughout the region. Introduced species, first reported just east of the region in 1935. Species is likely causing widespread decline in native hole nesting species because of competition for nesting sites.
County Records: [A, B, C, D, F, G, HL, HM, K, L, M, P]

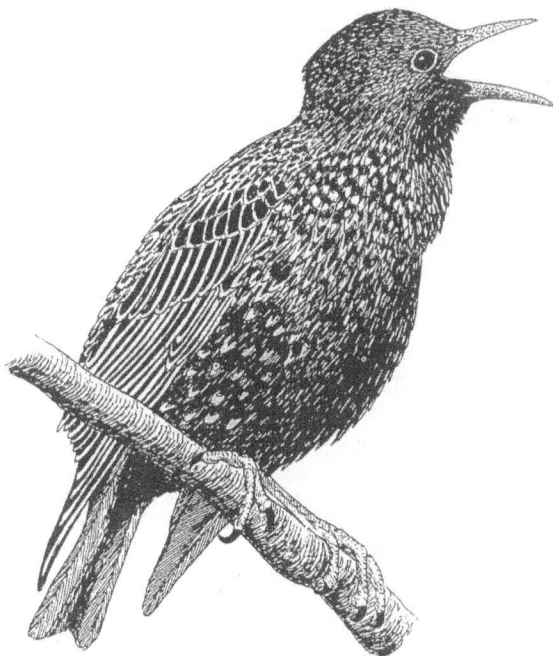

European Starling, adult male

Family Motacillidae: Pipits

American Pipit, *Anthus rubescens*
Uncommon spring and fall migrant and rare summer and winter visitor throughout the region, migrants to be expected mid-March to late May, and early September to mid-November.
County Records: [A, B, C, D, F, G, HL, HM, K, L, M, P]

Sprague's Pipit, *Anthus spragueii*
Uncommon spring and fall migrant, migrants to be expected late March to late May, and mid-September to late October. ABC Yellow-List species.
County Records: [A, C, HL, K, L, P]

Family Bombycillidae: Waxwings

Bohemian Waxwing, *Bombycilla garrulus*
Occasional winter visitor, to be expected early November to mid-April.
County Records: [A, L]

Sprague's Pipit, adult

Cedar Waxwing, adult

Cedar Waxwing, *Bombycilla cedrorum*
Common spring and fall migrant, year-round resident, and erratic, semi-regular breeder throughout the region, Migrants to be expected early March to late April, and late August to early November, plus non-breeding summer flocks.
County Records: [A, B, C, D, F, G, HL, HM, K, L, M, P]

Family Calcariidae: Longspurs and Snow Buntings

Lapland Longspur, *Calcarius lapponicus*
Uncommon winter visitor throughout the region, to be expected mid-October to mid-April. Very large winter aggregations have been reported in the region. *C. l. alascensis* and *C. l. lapponicus* occur in region (AOU, 1998, Sharpe et al., 2001).
County Records: [A, B, C, D, F, G, HL, HM, L, M, P]

Chestnut-collared Longspur, *Calcarius ornatus*
Rare winter visitor, to be expected late September to mid-April.
Tier 2 At-Risk species and ABC Yellow-List species.
County Records: [A, C, HL, HM, L]

Smith's Longspur, *Calcarius pictus*
Very rare winter visitor to be expected early December to late January. ABC Yellow-List species.
County Records: [A, C]

McCown's Longspur, *Calcarius mcownii*
Rare spring, fall and winter visitor, to be expected late October to late March. Tier 1 A Tier 2 Nebraska At-Risk species
County Records: [A, C, L]

Snow Bunting, *Plectrophenax nivalis*
Rare fall and winter visitor, to be expected mid-November to late February.
County Records: [A, C, P]

Family Parulidae: Wood Warblers

Ovenbird, *Seiurus aurocapillus*
Uncommon spring and fall migrant and possible rare summer visitor, to be expected during May and September. No evidence of regional nesting exists, but Lingle (1994) reported regional May, June and August records.
County Records: [A, B, C, F, HL, L]

Worm-eating Warbler, *Helmintheros vermivorum*
Very rare, with a single record. One was observed at Lange WPA, Clay County, on May 1, 1999 (Jorgensen, 2012).
County Records: [C]

Louisiana Waterthrush, *Parksia motacilla*
Rare migrant, to be expected in May and September. Reported regionally in May (Lingle, 1994). There is a specimen record for May 4, 1920, and a sight record for August 30, 1932, both in Lincoln County (Tout, 1947). Jorgensen (2012) found no reliable occurrence records for the Rainwater Basin. Tier 2 Nebraska At-Risk species.
County Records: [HL, L].

Northern Waterthrush, *Parksia noveboracensis*
Uncommon migrant, to be expected in May, and from late August to late September, mainly in east. Reported regionally in May (Lingle, 1994).
County Records: [A, B, C, HL]

Golden-winged Warbler, *Vermivora chrysoptera*
Very rare spring migrant, to be expected in May and September in the east. Reported regionally in May, but with no specific county attribution (Lingle, 1994), and in Adams County May 17, 1954 (NBR 22:64). ABC Red-List species.
County Records: [A]

Blue-winged Warbler, *Vermivora pinus*
Very rare migrant, to be expected in May and August–September (east). Reported in Adams County, May 15, 1967 (NBR 35:15), and May 19, 1962 (NBR 30:68). ABC Yellow-List species.
County Records: [A]

Black-and-White Warbler, *Mniotilta varia*
Uncommon spring and fall migrant, to be expected in May and September. No evidence of regional nesting, which sometimes occurs in northern Nebraska. Tier 2 At-Risk species.
County Records: [A, B, C, HL, L, M, P].

Prothonotary Warbler, *Protonotaria citrea*
Rare spring migrant, Reported May 24, 1968, from Lincoln County (Sharpe et al., 2001). There are reports from Adams County for May 12, 1956 (NBR 24:73), and September 4–12, 1973 (NBR 42:37). ABC Yellow-List species.
County Records: [A, L]

Swainson's Warbler, *Lymnothlypis swainsonii*
Very rare spring migrant. A specimen was reportedly collected April 9, 1905 at Kearney (Sharpe et al., 2001). ABC Yellow-List species.
County Records: [B]

Tennessee Warbler, *Oreothlypis peregrina*
Uncommon spring migrant throughout the region,, to be expected from early to late May, and from early September to early October. A bird with an active brood patch, captured during the spring of

1986 in Lincoln County, was suggestive of local (but highly extra-limital) nesting.
County Records: [A, B, C, G, HM, K, L, M, P]

Orange-crowned Warbler, *Oreothlypis celata*
Common spring and fall migrant throughout the region, to be expected from late April to mid-May, and from mid-September to mid-October. Reported regionally in April, May, September and October (Lingle, 1994).
County Records: [A, B, C, F, HL, HM, K, L, M, P]

Nashville Warbler, *Oreothlypis ruficapilla*
Uncommon spring an fall migrant throughout the region, to be expected from early to mid-May, and from mid-September to early October. Reported regionally in May, September and October (Lingle, 1994).
County Records: [A, B, C, F, G, HL, HM, K, L, M, P]

Virginia's Warbler, *Oreothlypis virginiae*
Very rare spring migrant. Reported at North Platte Fish Hatchery, Lincoln County April 26 and 29, 1964 (Shickley, 1968; Sharpe et al., 2001). ABC Yellow-List species.
County Records: [L]

Connecticut Warbler, *Oporonis agilis*
Rare spring migrant. Reported from Hamilton County (NBR 23:78) and Lincoln County, April 26, 1970 (Sharpe et al., 2001).
County Records: [C, HM, L]

MacGillivray's Warbler, *Oporornis tolmiei*
Rare migrant, mostly in western areas, to be expected in May and September. Reported from Clay County, May 15, 1999 (Sharpe et al., 2001), and observed during spring of 1935 from Lincoln County (Tout, 1947). There are spring records from Adams County (Jorgensen, 2012).
County Records: [A, C, L]

Mourning Warbler, *Oporonis philadelphia*
Rare spring and fall migrant, to be expected from mid- to late May, and early September to early October (east). Reported from Lincoln County September 29 to October 3, 1975 (Sharpe et al., 2001).
County Records: [A, C, L]

Kentucky Warbler, *Geothlypis formosa*
Rare spring migrant. Reported from Adams County on May 15, 1951 (NBR 19:74) and May 16, 1950 (NBR 18:78–79). ABC Yellow-List species.
County Records: [A, L]

Common Yellowthroat, *Geothlypis trichas*
Very common migrant and local summer resident throughout the region, to be expected from early May to mid-September. Reported regionally for all months from April to September by Lingle (1994). Considered a common breeder in the Rainwater Basin by Jorgensen (2012). Mollhoff (2001) confirmed nesting in Adams, Buffalo, Clay, Hall, Hamilton and Lincoln counties.
County Records: [A, B, C, D, F, G, HL, HM, K, L, M, P]

American Redstart, *Setophaga ruticilla*
Uncommon migrant and probable highly local summer resident throughout the region, from early May to mid-September. Breeding was reported for Adams County in 1967 (Bennett, 1968). Reported regionally during May and August by Lingle (1994).
County Records: [A, B, C, D, HL, HM, K, L, M, P]

Cape May Warbler, *Setophaga tigrina*
Rare migrant, to be expected in May and September. Reported regionally during May by Lingle (1994). Jorgensen (2012) mentions Rainwater Basin records from May 10–22, from Adams County (NBR 3:95; 49:51)
County Records: [A]

Cerulean Warbler, *Setophaga cerulea*
Rare spring and fall migrant Reported on August 27, 1978 in Lincoln County, and on September 2, 1973 and September 3, 1975 in Adams County (Sharpe et al., 2001). This eastern species has nearly been extirpated from Nebraska. Tier 2 At-Risk species and ABC Yellow-List species.
County Records: [A, L]

Northern Parula, *Parula americana*
Rare spring migrant. Reported from North Platte, Lincoln County May 19, 1935, and May 1, 1937 (Tout, 1947). There is a May 6 record from Lange WPA, Clay County, four records from Adams County,

from April 25–May 10, and two fall records for September 2 and 3 (Jorgensen, 2012).
County Records: [A, L]

Magnolia Warbler, *Setophaga magnolia*
Rare spring and fall migrant, to be expected from mid- to late May, and early September to early October. Reported regionally during May by Lingle (1994). Reported from Adams County (NBR 30:47; Jorgensen, 2012), Jorgensen noted that spring records from the Rainwater Basin were from May 7–23, and mentioned a single fall record, for September 24.
County Records: [A, HL]

Bay-breasted Warbler, *Setophaga castanea*
Rare spring and fall migrant, to be expected from May and September. Reported regionally during May by Lingle (1994), but county attributions are limited to Adams County (Jorgensen 2012). ABC Yellow-List species.
County Records: [A]

Blackburnian Warbler, *Setophaga fusca*
Rare spring and fall migrant, to be expected in May and from early September to early October. Reported regionally during May by Lingle (1994).
County Records: [A, C, HL, HM]

Yellow Warbler, *Setophaga petechia*
Very common migrant and local summer resident throughout the region, to be expected from early May to early September. Reported regionally for all months from May to September by Lingle (1994).
County Records: [A, B, C, D, F, G, HL, HM, K, L, M, P]

Chestnut-sided Warbler, *Setophaga pensylvanica*
Rare spring and fall migrant, to be expected mid- to late May, and early to late September, Reported regionally during May by Lingle (1994).
County Records: [A, B, HL]

Blackpoll Warbler, *Setophaga striata*
Uncommon spring and probable fall migrant, to be expected May and September. Reported regionally during May by Lingle (1994).

Jorgensen (2012) found spring Rainwater Basin reports and fall reports, but considered all these fall reports as hypothetical, owing to this species' similarity to fall-plumage Bay-breasted Warblers.
County Records: [A, B, C, HL, HM, P]

Black-throated Blue Warbler, *Setophaga caerulescens.*
Rare spring and fall migrant, Reported regionally during May by Lingle (1994), and on October 3, 1934 in Lincoln County by Tout (1947). There is a record from Adams County, May 17, 1978 (NBR 46:80).
County Records: [A, L]

Palm Warbler, *Setophaga palmarum*
Rare spring and fall migrant, to be expected May and September. Reported regionally during May by Lingle (1994).
County Records: [B, C, D, HL, L]

Pine Warbler, *Setophaga pinus*
Rare spring and fall migrant. Reported from Lincoln County, December 24, 2007, which is a very questionable date for any warbler to be present in Nebraska, except for the Yellow-rumped Warbler.
County Records: [HL, K, L]

Yellow-rumped Warbler, *Setophaga coronata*
Very common (*S. c. coronata*) to occasional (*S. c. auduboni*) spring and fall migrant throughout the region, to be expected from late April to May, and from September to late October. Reported regionally during April, May, September and October by Lingle (1994).
County Records: [A, B, C, D, F, G, HL, HM, K, L, M, P]

Yellow-throated Warbler, *Setophaga dominica*
Probably a very rare spring and fall migrant. Reported regionally during May by Lingle (1994), Tier 2 At-Risk species.
County Records: [A, L]

Prairie Warbler, *Setophaga discolor*
Very rare spring migrant. One was reported from Brady, Lincoln County on May 8, 2009 and there is an Adams County report for May 12, 1962 (NBR 30:54). ABC Yellow-List species
County Records: [A, HL, L]

[**Black-throated Gray Warbler**, *Setophaga nigrescens*]
Very rare spring migrant. This southwestern warbler was reported without documentation from Adams County on May 11, 1971 (Sharpe et al., 2001).
County Records: [A]

Townsend's Warbler, *Setophaga townsendi*
Very rare spring and fall migrant. Spring records include April 27, 1978, May 5, 1978 and May 8, 1973 from Buffalo County (Sharpe et al., 2001), and May 11, 1971 from Adams County (NBR 39:54). Fall records from Lincoln County include July 14, 1965, and August 17, 1966 (Shickley, 1968; Sharpe et al., 2001).
County Records: [A, B, C, L]

Black-throated Green Warbler, *Setophaga virens*
Rare spring and fall migrant. Lingle (1994) reported regional May, June and October records. It has been reported from Adams County on several occasions (NBR 5:19; 5:63; 6:18; 34:32, 30:54) and from Clay County (Jorgensen, 2012).
County Records: [A, C, HL]

Canada Warbler, *Cardellina canadensis*
Uncommon spring and rare fall migrant, to be expected May and September. County records include May 17, 1916 in Clay County, May 20, 1935 in Adams County, and May 25, 1957, in Hall County (Sharpe et al., 2001).
County Records: [A, C, H]

Wilson's Warbler, *Cardellina pusilla*
Uncommon spring and fall migrant, to be expected May and September (mainly east). Reported regionally during May, July, August and September by Lingle (1994). Specific county records include April 21, 1972 in Adams County, and June 7, 1998 in Phelps County (Sharpe et al., 2001).
County Records: [A, B, C, F, HL, K, L, M, P]

Yellow-breasted Chat, *Icteria virens*
Common migrant and previously a local summer resident, to be expected mid-May to early September. This species has been in marked decline, especially in eastern Nebraska, from which it has largely disappeared. Reported regionally during May by Lin-

gle (1994). Jorgensen (2012) reported a summer presence in Adams County during 1976.
County Records: [A, B, C, D, F, G, HL, M, P]

Family Emberizidae: Towhees and Sparrows

Spotted Towhee, *Pipilo maculatus*
Common migrant and local summer resident in the west, to be expected late April to mid-October, and an uncommon winter resident. The taxonomically inclusive "Rufous-sided" Towhee (which from the late 1950's until 1995 consisted of the Spotted and Eastern towhees) has been reported regionally for all months except March (Lingle, 1994). Breeding by this taxon has been reported for Lincoln County (Tout, 1947), and interbreeding farther east along the Platte Valley with the Eastern Towhee is widespread (Short, 1961). Most towhees nesting in Lincoln County are Spotteds, although hybrids also occur (NBR 73: 110). In the Gibbon, Buffalo County area, three times as many Spotted as Eastern Towhees were seen, plus many hybrids (NBR 71:124).
County Records: [A, B, C, D, F, G, HL, HM, K, L, M, P]

Eastern Towhee, *Pipilo erythrophthalmus*
Common migrant and local summer resident to be expected late April to mid-October (east). Interbreeding with the probably conspecific Spotted Towhee is widespread in the central Platte Valley (Short, 1961). Eastern Towhees probably breed in the Platte Valley west to about Hall County (NBR 64:124). Extensive hybridization in the towhees of the central Platte Valley was documented by Sibley and West (1959). Jorgensen (2012) judged both towhees to be only spring and fall migrants in the Rainwater Basin.
County Records: [A, B, C, D, F, G, HL, K, L, M, P]

American Tree Sparrow, *Spizella arborea*
Very common wintering migrant throughout the region, to be expected late October to early April. Reported regionally for all months from October to May (Lingle, 1994).
County Records: [A, B, C, D, F, G, HL, HM, K, L, M, P]

Chipping Sparrow, *Spizella passerinea*
Very common migrant and local summer resident throughout the region, to be expected late April to early October, but rarely re-

maining until early winter. Reported regionally for all months from March to December (Lingle, 1994). Common summer resident in the Rainwater Basin by Jorgensen (2012). Confirmed nesting in Clay, Hall and Lincoln counties was reported by Mollhoff (2001).
County Records: [A, B, C, D, F, G, HL, HM, K, L, M, P]

Clay-colored Sparrow, *Spizella pallida*
Very common migrant throughout the region, to be expected early to mid-May, and mid-September to early October. Reported regionally from April to May and from September to October (Lingle, 1994). An undocumented report breeding in 1973 was indicated for Hall County (Bennett, 1974).
County Records: [A, B, C, D, F, G, HL, HM, K, L, M, P]

Field Sparrow, *Spizella pusilla*
Very common migrant and local summer resident throughout the region, to be expected mid-April to early October. Reported regionally for all months from February to December, excepting March and November (Lingle, 1994). Jorgensen (2012) found no records of confirmed breeding in the Rainwater Basin, but Mollhoff (2001) reported cases of probable breeding in Dawson, Frontier, Hall, Hamilton, Kearney and Lincoln counties.
County Records: [A, B, C, D, F, G, HL, HM, K, L, M, P]

Vesper Sparrow, *Pooecetes gramineus*
Very common migrant throughout the region and possible local summer resident, to be expected mid-April to early October. Reported regionally from March to May and from August to November, plus January (Lingle, 1994). Jorgensen (2012) found no records of confirmed nesting in the Rainwater Basin, and Mollhoff (2001) reported only one case of probable regional nesting in Dawson County.
County Records: [A, B, C, D, F, G, HL, HM, K, L, M, P]

Lark Sparrow, *Chondestes grammacus*
Very common migrant and summer resident throughout the region (mainly westerly), to be expected early May to early September. Reported regionally from May to August (Lingle, 1994). Breeding has been reported from Adams County (NBR 34:43), as well as from Buffalo, Frontier and Lincoln counties (Mollhoff, 2001). Jorgensen (2012) found no records of confirmed breeding in the Rainwater Basin.
County Records: [A, B, C, D, F, G, HL, HM, K, L, M, P]

Lark Sparrow, adult male

Lark Bunting, *Calamospiza melanocorys*
Common migrant and summer resident throughout the region, to be expected early May to late August (west). Reported regionally for all months from April to October (Lingle, 1994). Mollhoff (2001) reported cases of confirmed regional breeding during the middle and later 1980's in Adams, Dawson, and Lincoln counties. Jorgensen (2012) found no records of confirmed breeding in the Rainwater Basin since 1965. ABC Yellow-List species.
County Records: [A, B, C, D, F, G, HL, K, L, M, P]

Savannah Sparrow, *Passerculus sandwichensis*
Very common migrant throughout the region, to be expected late April to mid-May, and mid-September to mid-October. Reported regionally from March to May and from September to November (Lingle, 1994).
County Records: [A, B, C, D, F, G, HL, HM, K, L, M, P]

Grasshopper Sparrow, *Ammodramus savannarum*
Very common migrant and uncommon summer resident throughout the region, to be expected early May to early September. Reported regionally for all months from April to October (Lingle, 1994). Mollhoff (2001) reported cases of confirmed regional breed-

ing during the middle and later 1980s in Adams, Dawson, and Lincoln counties. Jorgensen (2012) mentioned breeding at Hultine WPA, Clay County in 1999.
County Records: [A, B, C, D, F, G, HL, HM, K, L, M, P]

Baird's Sparrow, *Ammodramus bairdii*
Probably a very rare migrant, but the only published regional record is a sighting September 23, 1969, from Adams County (NBR 38:38). ABC Red-List species.
County Records: [A, B]

Henslow's Sparrow, *Ammodramus henslowii*
Rare migrant and very local summer resident (east), to be expected from late April to late September, Reported regionally during May and Sept (Lingle, 1994). This rare tallgrass sparrow is known to have nested at the Crane Trust, Buffalo County, starting at least as early as 1995, and during three of four years from 2004–2007 (NBR 75:84). Singing has also been heard at Harvard and Lange WPAs, Clay County (Jorgensen, 2012). Tier 2 At-Risk species and ABC Red-List species.
County Records: [A, B, C, HL]

Le Conte's Sparrow, *Ammodramus leconteii*
Common migrant, to be expected from late April to early May, and late September to late October. Reported regionally from April to May and from September to November (Lingle, 1994).
County Records: [A, C, HL, HM]

Nelson's Sparrow, *Ammodramus nelsoni*
Rare migrant, to be expected from May and October. Reported regionally in March, May and October (Lingle, 1994). It has been reported during October at Harvard, Theesen, and North Hultine WPAs, Clay County (Jorgensen, 2012). ABC Yellow-List species. Previously called the Nelson's Sharp-tailed Sparrow.
County Records: [C, HL]

Fox Sparrow, *Passerella iliaca*
Common migrant, to be expected from late March to mid-April, and mid-October to mid-November, Reported regionally in April, May and November (Lingle, 1994), plus a few January records (NBR 41:64; 47:54).
County Records: [A, B, C, D, F, G, HL, HM, K, L, M, P]

Song Sparrow, *Melospiza melodia*
Very common migrant and local summer resident, to be expected from early April to late December, sometimes overwintering during mild winters. Reported regionally in every month (Lingle, 1994). The species has become a more common breeder in south-central Nebraska (Jorgensen, 2012).
County Records: [A, B, C, D, F, G, HL, HM, K, L, M, P]

Lincoln's Sparrow, *Melospiza lincolnii*
Very common migrant, to be expected from late April to mid-May, and mid-September to mid-October. Reported regionally from March to May and from September to December, excepting November (Lingle, 1994).
County Records: [A, B, C, D, F, G, HL, HM, K, L, M, P]

Swamp Sparrow, *Melospiza georgiana*
Uncommon migrant throughout the region, and highly local summer resident, to be expected from late April to late October. Reported regionally from April to May, and from September to October, plus a January record (Lingle, 1994). Marshes on the south side of the Platte River in Hall County are near the southern limits of the species' Nebraska breeding range (NBR 76: 110). Observations at Funk WPA, Phelps County, July 11, 2002, represent one of a few Nebraska breeding populations located well south of the Platte River (NBR 70:117). Jorgensen (2012) found no summer records more recent than July 2009, when a singing male was heard at Harvard WPA, Clay County. Tier 2 At-Risk species.
County Records: [A, B, C, HL, HM, K, L, M, P]

White-throated Sparrow, *Zonotrichia albicollis*
Very common spring and fall migrant throughout the region, sometimes overwintering, to be expected from early October to mid-May. Reported regionally from April to May and from October to December (Lingle, 1994).
County Records: [A, B, C, D, F, G, HL, HM, K, L, M, P]

Harris's Sparrow, *Zonotrichia querula*
Very common wintering migrant throughout the region, to be expected from mid-October to mid-May. Reported regionally for all months from October to May (Lingle, 1994).
County Records: [A, B, C, D, F, G, HL, HM, K, L, M, P]

White-crowned Sparrow, *Zonotrichia leucophrys*
Very common wintering migrant throughout the region, to be expected from early October to mid-May. Reported regionally for all months from September to June, excepting March (Lingle, 1994).
County Records: [A, B, C, D, F, G, HL, HM, K, L, M, P]

Dark-eyed Junco, *Junco hyemalis*
Very common wintering migrant throughout the region, to be expected from early October to late March. Reported regionally for all months from September to June (Lingle, 1994). The several races occurring in the state (*aikeni, cismontanus, hyemalis, mearnsi* and *montanus*) are variably uncommon to abundant. Tier 2 At-Risk species.
County Records: [A, B, C, D, F, G, HL, HM, K, L, M, P]

Family Cardinalidae: Cardinals, Tanagers, and Grosbeaks

Summer Tanager, *Piranga rubra*
Rare migrant, to be expected from May and September. Reported regionally in April (Lingle, 1994). A case of extralimital historic breeding was reported for Lincoln County in 1938 (Tout, 1939).
County Records: [HL, L]

Scarlet Tanager, *Piranga olivacea*
Uncommon migrant and rare summer resident, to be expected from mid-May to late August. Reported regionally in June (Lingle, 1994). There are records from Adams County (NBR 15:68) and Clay County (NBR 3:65).
County Records: [A, C, HL]

Northern Cardinal, *Cardinalis cardinalis*
Common resident throughout the region, especially in towns, More common eastwardly, but still expanding westward. Reported regionally in every month (Lingle, 1994).
County Records: [A, B, C, D, F, G, HL, HM, K, L, M, P]

Rose-breasted Grosbeak, *Pheucticus ludovicianus*
Very common migrant throughout the region, and local summer resident, to be expected from early May to mid-September (mainly in east). Reported regionally for all months from March to September (Lingle, 1994). There are confirmed breeding records from Ad-

ams County, and probable breedings during the 1980s from Buffalo, Clay, Hall, Hamilton and Merrick counties (Mollhoff, 2001).
County Records: [A, B, C, D, F, G, HL, HM, K, L, M, P]

Black-headed Grosbeak, *Pheucticus melanocephalus*
Common (west) to occasional (east) migrant and highly local summer resident, to be expected from mid-May to late August. Reported regionally for all months from April to July (Lingle, 1994). Gibbon, Buffalo County is the most eastern regular breeding site for Black-headed Grosbeaks in the Platte Valley, and Rose-breasted and Black-headed Grosbeaks are in contact there (NBR 69:136; 76:110). West (1962) documented extensive hybridization among the grosbeaks of the Platte Valley. Probable breeding records during the 1980's include Dawson, Gosper, Lincoln and Phelps counties (Mollhoff, 2001).
County Records: [A, B, D, F, G, HL, K, L, M, P]

Blue Grosbeak, *Passerina caerulea*
Very common migrant throughout the region (mainly westwardly), and local summer resident, to be expected from mid-May to late August. Reported regionally for all months from May to August (Lingle, 1994). There was a confirmed breeding record during the 1980's from Kearney County, and probable breedings from Gosper, Hamilton, Lincoln and Phelps counties (Mollhoff, 2001).
County Records: [A, B, C, D, F, G, HL, K, L, M, P]

Lazuli Bunting, *Passerina amoena*
Uncommon (west) to rare (east) migrant and highly local summer resident, to be expected from mid-May to late August. Reported regionally in May (Lingle, 1994). Possible breeding was reported for Lincoln County (Mollhoff, 2001). Extensive hybridization between Indigo Buntings versus Lazuli Buntings in the Platte Valley was documented by Sibley and Short (1959).
County Records: [A, B, C, K, L]

Indigo Bunting, *Passerina cyanea*
Very common migrant and local summer resident, to be expected from mid-May to late August (mainly in east and central). Reported regionally for all months from May to August (Lingle, 1994). Extensive hybridization between Indigo Buntings versus Lazuli Buntings in the Platte Valley was documented by Sibley and Short (1959).
County Records: [A, B, C, D, F, G, HL, HM, K, L, M, P]

Painted Bunting, *Passerina ciris*
　　Rare migrant. Reported from Adams County, May 19, 1962 (NBR
　　30:70), Lincoln County, June 4, 1973 (NBR 41:67–68, 1973); and from
　　Buffalo County, May 14, 2008. Regionally reported for November
　　(Lingle, 1994). ABC Yellow-List species
　　County Records: [A, B, C, HL, L]

Dickcissel, *Spiza americana*
　　Very common migrant throughout the region, and local summer
　　resident, to be expected from mid-May to late August. Reported re-
　　gionally for all months from May to October (Lingle, 1994).
　　County Records: [A, B, C, D, F, G, HL, HM, K, L, M, P]

Family Icteridae: Meadowlarks, Blackbirds, and Orioles

Bobolink, *Dolichonyx oryzivorus*
　　Very common migrant throughout the region, and local summer
　　resident, throughout the region, to be expected from mid-May to
　　mid-August. Reported regionally from April to August (Lingle,
　　1994). Jorgensen (2012) mentioned several locations in Clay County,
　　where singing males have been reported during summer.
　　County Records: [A, B, C, D, F, G, HL, HM, K, L, M, P]

Red-winged Blackbird, *Agelaius phoeniceus*
　　Very common migrant and abundant summer resident throughout
　　the region, to be expected from early March to late November. Re-
　　ported regionally for every month (Lingle, 1994).
　　County Records: [A, B, C, D, F, G, HL, HM, K, L, M, P]

Eastern Meadowlark, *Sturnella magna*
　　Common migrant throughout the region, and local summer resi-
　　dent (mainly in lowland habitats toward the east), to be expected
　　from early April to mid-October. Reported regionally for all months
　　from March to July (Lingle, 1994). Breeding has been confirmed for
　　Lincoln County (Mollhoff, 2001). Jorgensen (202) found this species
　　to be present only during spring, with a single summer record in
　　Clay County.
　　County Records: [A, B, C, F, HL, HM, K, L, M, P]

Western Meadowlark, *Sturnella neglecta*
　　Very common migrant and abundant summer resident throughout

Red-winged Blackbird, adult male

the region, to be expected from early March to late October. Reported regionally for every month (Lingle, 1994).
County Records: [A, B, C, D, F, G, HL, HM, K, L, M, P]

Yellow-headed Blackbird, *Xanthocephalus xanthocephalus*
Very common migrant and scattered summer resident throughout the region, to be expected from mid-April to mid-September. Reported regionally for all months from March to November (Lingle, 1994). Mollhoff (2001) reported confirmed breeding records for Adams, Clay, Dawson, Lincoln and Phelps counties. Jorgensen (2012) reported that this species is not as common in the Rainwater Basin as previously, and has recently been limited to larger wetlands during wet years.
County Records: [A, B, C, D, F, G, HL, HM, K, L, M, P]

Rusty Blackbird, *Euphagus carolinus*
Occasional spring and fall migrant and winter visitor, to be expected from late March to mid-April, and early November to late December, frequently overwintering. Reported regionally for all months from October to March, excepting January (Lingle, 1994). A supposed 1982 nesting in Hall County is not likely, and Jorgensen (2012) judged that all June records in the Rainwater Basin were the result of misidentification.
County Records: [A, B, C, HL, HM, K, L, M, P]

Brewer's Blackbird, *Euphagus cyanocephalus*
Uncommon migrant and highly local summer resident, to be expected from mid-April to early November (west). Reported regionally for all months from September to April, excluding February (Lingle, 1994). Probable breeding in Hall County was reported in 1972 and 1974 (Bennett, 1973, 1975) and was later confirmed for Clay County (Mollhoff, 2001). Tier 2 At-Risk species.
County Records: [A, B, C, D, F, G, HL, HM, K, L, M, P]

Common Grackle, *Quiscalus quiscula*
Very common migrant and abundant summer resident throughout the region, to be expected from late March to late October. Reported regionally for every month (Lingle, 1994).
County Records: [A, B, C, D, HL, K, L, M, P]

Great-tailed Grackle, *Quiscalus mexicanus*
Increasingly common migrant and local summer resident, to be expected from April to September. Regional breeding first began in the region in 1977, in Adams County (NBR 45:18), and was followed in later years by breedings at Funk WPA, Phelps County, and at Smith and Kissinger Basin WPAs, Clay County (Jorgensen, 2012).
County Records: [A, B, C, D, F, HL, K, L, M, P]

Brown-headed Cowbird, *Molothrus ater*
Very common migrant and uncommon summer resident throughout the region, to be expected from mid-April to early October. Reported regionally for all months from March to December, excluding November (Lingle, 1994).
County Records: [A, B, C, D, F, G, HL, HM, K, L, M, P]

Orchard Oriole, *Icterus spurius*
Very common migrant and uncommon summer resident through-out the region, to be expected from mid-May to late August. Reported regionally for all months from May to August (Lingle, 1994). County Records: [A, B, C, D, F, G, HL, HM, K, L, M, P]

Bullock's Oriole, *Icterus bullockii*
Occasional migrant and local summer resident (in west), to be expected from early May to early September. Two birds with brood patches were in banded in Dawson County 2001–2002 and possible breeding has been reported for Lincoln County (Mollhoff, 2001). There is a 1915 specimen record for Kearney, Buffalo County (Sharpe et al., 2001). The incidence of Bullock's x Baltimore oriole hybrids during the 1950's was documented in the central Platte River Valley by Sibley and Short (1964). Jorgensen (2012) found a single Rainwater Basin record, from Adams County, May 13, 1939. County Records: [A, B, D, L]

Baltimore Oriole, *Icterus galbula*
Very common migrant and common summer resident through-out the region, to be expected from early May to early September. Records from 1973 to 1998 for the taxonomically more inclusive "Northern" Oriole (Baltimore and Bullock's orioles) extend regionally from May to September (Lingle, 1994). Extensive hybridization between these orioles in the Platte River Valley and Great Plains was first documented by Sibley and Short (1964), and later studied by Rising (1970). Rising determined that the incidence of hybridization appeared to be declining, suggesting that the two were increasingly responding to each other as if they were separate species, rather than as sometimes interbreeding subspecies. County Records: [A, B, C, D, F, G, HL, HM, K, L, M, P]

Scott's Oriole, *Icterus parisorum*
Very rare migrant. Reported from Kearney on June 29, 2004, the fifth reported state record (NBR 72:93). There is also an undocumented record for Adams County, in May, 1951 (NBR 19:75). County Records: [A, B, HL]

Family Fringillidae: Finches

[**Gray-crowned Rosy-Finch**, *Leucosticte tephrocotis*]
Hypothetical. Reported in Lincoln County, December 7, 1910 (Tout, 1947). A report for Hall County, Nov 7, 1993, was not accepted by the NOU Records Committee (Sharpe et al., 2001).
County Records: [HL]

Pine Grosbeak, *Pinicola enucleator*
Rare wintering migrant, to be expected from late November to mid-March. Reported regionally in December (Lingle 1994),
County Records: [A]

Purple Finch, *Carpodacus purpureus*
Periodically common wintering migrant, to be expected from late October to late April. Reported regionally for all months from November to January (Lingle, 1994). Jorgensen (2012) found no Rainwater Basin records more recent than 1986.
County Records: [A, B, C, HL, K, L, M, P]

Cassin's Finch, *Carpodacus cassinii*
Very rare wintering migrant, to be expected from October to mid-April. An individual was banded at North Platte, Lincoln County, November 22, 1936 (Tout, 1947), and there is a report for January 8, 1958 from Adams County (NBR 26:65).
County Records: [A, L]

House Finch, *Carpodacus mexicanus*
Very common local resident throughout the region, especially in towns and villages. Reported regionally in every month (Lingle, 1994). It current presence in the Platte River Valley is the result of expansion from both eastern and western populations that reached the central Platte Valley during the late 1960's, when it was first reported from Adams County (Jorgensen, 2012). The region's first breeding reports were from Hall County in 1983 (Sharpe et al., 2001).
County Records: [A, B, C, D, F, G, HL, HM, K, L, M, P]

Red Crossbill, *Loxia curvirostra*
Periodically uncommon and irruptive wintering migrant, to be expected from mid-November to early April. Reported regionally for

all months from December to April (Lingle, 1994). A reported 1966 breeding at Hastings, Adams County, involved finding two adults and a young bird in late July, but the juvenile might have fledged elsewhere (Sharpe et al., 2001).
County Records: [A, B, HL, K, L, M, P]

White-winged Crossbill, *Loxia leucoptera*
Rare overwintering migrant, to be expected from October to April. Reported regionally in January (Lingle 1994), and as late as May (Jorgensen, 2012).
County Records: [A, H, L]

Common Redpoll, *Acanthus flammea*
Rare wintering migrant, to be expected from late November to mid-March. Reported regionally in January and April (Lingle, 1994).
County Records: [B, HL]

Pine Siskin, *Spinus pinus*
Common wintering migrant, to be expected from mid-October to mid-May; also an irregular and highly local summer resident. Reported regionally for all months from September to May (Lingle, 1994). There are several nesting records for Adams County (Jorgensen, 2012). Tier 2 At-Risk species.
County Records: [A, B, C, D, F, G, HL, HM, K, L, M, P]

Lesser Goldfinch, *Carduelis psaltria*
Very rare migrant. Questionably reported from Buffalo County in November 1996, and from Hall County, November 7, 1993 (Sharpe et al., 2001).
County Records: [B, HL]

American Goldfinch, *Carduelis tristis*
Very common local resident throughout the region. Reported regionally in every month (Lingle, 1994).
County Records: [A, B, C, D, F, G, HL, HM, K, L, M, P]

Evening Grosbeak, *Coccothraustes vespertinus*
Periodically uncommon overwintering migrant, to be expected from early November to late April. Reported regionally for all months from November to April (Lingle, 1994).
County Records: [A, B, HL]

House Sparrow, adult male

Family Fringillidae: Old World Sparrows

House Sparrow, *Passer domesticus*
   Abundant introduced resident throughout the region, mainly in towns and around farms. Reported regionally in every month (Lingle, 1994). House sparrows were reportedly purposefully introduced into the Platte Valley by farmers during the late 1800's to help control grasshoppers, but they had the opposite agricultural effect, by being mainly grain-eaters rather than insect-eaters.
   County Records: [A, B, C, D, F, G, HL, HM, K, L, M, P]

Locations of Birding Sites in the Central Platte River Valley

Latitude/longitude coordinates are included for nearly all sites, more detailed descriptions and mapped directions of the sites can also be found in *A Nebraska Bird-finding Guide* (Johnsgard, 2012) (http://digitalcommons.unl.edu/biosciornithology/51). Additional site information is available online at the Nebraska Birding Trails website: (http://nebraskabirdingtrails.com).

ADAMS COUNTY

Kenesaw WPA. Area: 161 acres. Located 0.5 mile east and 0.5 mile south of Kenesaw. Lat/Long 40.602 N/98.645 W.

Weseman WPA. Area: 80 acres. Located 9 miles west and 4 miles south of Hastings, or 0.25 mile west of Assumption. Lat/Long 42.756 N/97.095 W.

BUFFALO COUNTY

Bassway Strip WMA. Area: 725 acres. Located 6 miles east of Kearney and west of 1-80 exit 279 south on U.S. Highway 10 on marked access road. Lat/Long 40.686 N/98.943 W.

Blue Hole WMA. Area: 530 acres. Located 2 miles south of Elm Creek. Lat/Long 40.683 N/99.327 W.

Rowe Sanctuary and Iain Nicolson Audubon Center. Area: *ca.* 2,000 acres. Located 2 miles south and 2 miles west of I-80 exit 285, on Elm Island Road. Lat/Long 40.178 N/98.531 W.

CLAY COUNTY

Bluewing WMA. Area: 160 acres. Located 4 miles west and 0.5 mile south of Edgar. Lat/Long 40.365 N/98.042 W.

Bulrush WMA. Area: 160 acres. Located 3.5 miles west of Edgar.
Lat/Long 40.392 N/98.076 W.

Eckhardt WPA. Area: 66 acres. Located 8 miles east and 4 miles south
of Clay Center.
Lat/Long 40.465 N/99.903 W.

Harms WPA. Area: 33 acres. Located 2 miles east and 2.5 miles south
of Clay Center.
Lat/Long 49.490 N/98.010 W.

Glenvil WPA. Area: 83 acres. Located 1.5 mile east and 1.5 mile south
of Glenvil.
Lat/Long 40.475 N/98.220 W.

Green Acres WPA. Area: 48 acres. Located 6 miles east and 4 miles
south of Clay Center, or 1.5 miles east and 6.5 miles north of Edgar.
Lat/Long 40.460 N/97.938 W.

Green Wing WMA. Area: 53 acres. Located 0.5 mile east and 3 miles
north of Ong.
Lat/Long 40.443 N/97.829 W.

Greenhead WMA. Area: 60 acres. Located 6 miles east and 6 miles
south of Clay Center.
Lat/Long 40.444 N/97.940 W.

Hansen WPA. Area: 296 acres. Located 0.25 mile west and 3.5 miles
north of Ong.
Lat/Long 40.449 N/97.848 W.

Harvard WPA. Area: 760 acres. Located 2 miles west of Harvard.
Lat/Long 40.614 N/98.181 W.

Hultine WPA. Area: 583 acres. Located 6 miles east of Harvard.
Lat/Long 40.628 N/97.973 W.

Kissinger Basin WMA. 490 acres. Located 0.5 mile north of Fairfield.
Lat/Long 40.445 N/98.099 W.

Lange WPA. Area: 56 acres. Located 0.25 mile east and 2 miles south of
Sutton.
Lat/Long 40.563 N/97.846 W.

Massie WPA. Area: 494 acres. Located 3 miles south and 0.5 mile east
of Clay Center.
Lat/Long 40.479 N/98.083 W.

Meadowlark WPA. Area: 80 acres. Located 3 miles east and 3 miles
south of Clay Center. Lat/Long 40.474 N/99.995 W.

Moger WPA. Area: 72 acres. Located 3 miles east and 2 miles south of Clay Center.
Lat/Long 40.483 N/97.899 W.

North Hultine WPA. Area: 240 acres. Located 6 miles east and 1 mile north of Harvard.
Lat/Long 40.499 N/97.714 W.

Shuck WPA. Area: 89 acres. Located 1 mile west and 5.5 miles south of Geneva.
Lat/Long 40.458 N/99.997 W.

Smith WPA. Area: 226 acres. Located 6 miles south and 3.5 miles east of Clay Center, or four miles north of Edgar.
Lat/Long 40.439 N/97.974 W

Theesen WPA. Area: 46 acres. Located 1.5 mile northwest of Glenville.
Lat/Long 40.512 N/98.273 W.

Verona WPA. Area: 160 acres. Located 4.5 miles east and 1.5 miles north of Clay Center.
Lat/Long 40.512 N/98.273 W.

White Front WMA. Area: 280 acres. Located 2 miles west and 1.5 mile north of Clay Center.
Lat/Long 40.550 N/-98.082 W.

DAWSON COUNTY

Bittern's Call WMA. Area: 80 acres. Located about 10 miles north of Lexington on Highway 2l. Lat/Long 40.842 N/99.846 W.

Cozad WMA. Area: 198 acres. Located 1 mile south and 1 mile east of Cozad.
Lat./Long. 40.837 N/99.981 W

Darr Strip WMA. Area: 976 acres. Located 1 mile south of Darr.
Lat/Long 40.810 N/99.899 W.

Dogwood WMA. Area: 402 acres. Located 4 miles west and 3 miles south of Overton.
Lat/Long 40.699 N/99.627 W.

East Willow Island WMA. Area: 37 acres. Located 3 miles west and 1 mile south of Cozad.
Lat/Long 40.863 N/100.036 W

J-2 Hydroplant. Located 6 miles south and west of Lexington on U.S. 283.
Lat/Long 40.414 N/99.490 W.

Johnson Lake SRA. Area: 2,229 acres. Located 7 miles south of Lexington on U.S. Highway 283.
Lat/Long 40.685 N/99.830 W.

West Cozad WMA. Area: 50 acres. Located 2 miles west and 1 mile south of Cozad
Lat/Long 40.685 N/99.830 W.

Willow Island WMA. Area 80 acres. Located five miles west of Cozad.
Lat/Long 40.876 N/100.063 W.

FRONTIER COUNTY

Red Willow Reservoir SRA/WMA. Area: 43 acres. Located 30 miles south of Curtis.
Lat/Long 40.372 N/100.711 W (WMA), 40.360 N/100.664 (SRA).

Medicine Creek Reservoir (Harry Strunk Lake) and Medicine Creek SRA/WMA. Area: SRA 1,768 acres, WMA 6,726 acres. Located 33 miles south of Stockville.
Lat/Long 40.437 N/100.265 (WMA), 40.379 N/100.212 (SRA).

GOSPER COUNTY

Elley Lagoon WPA. Area: 62 acres. Located 3 miles south and 2 miles west of Bertrand.
Lat/Long 40.486 N/99.684 W.

Elwood Reservoir WMA. Area: 2,200 acres. Located 2 miles north of Elwood.
Lat/Long 40.627 N/99.858 W.

Peterson WPA. Area: 1,154 acres. Located 1 mile west and 2 miles south of Bertrand.
Lat/Long 40.489 N/99.658 W.

Victor Lake WPA. Area: 238 acres. Located 4 miles north, 3 miles east, and 1 mile north or Bertrand.
Lat/Long 40.594 N/99.654 W.

HALL COUNTY

Crane Trust (previously The Platte River Whooping Crane Maintenance Trust). Located 2 miles south on Alda Road from I-80 Exit 305, turn east on Whooping Crane Drive, then continue east on Sandhill Crane Drive about 1 mile to headquarters.
Lat/Long 40.472 N/98.275 W.

Hannon WPA. Area: 659 acres. Located 1 mile east and 2 miles north of the I-80 Shelton Exit 292.
Lat/Long 40.752 N/98.697 W.

Loch Linda WMA. Area: 38 acres. Located 1 mile north, 2 miles east, 1 mile south over Interstate 80 and 2 miles east from Alda I-80 exit 305.
Lat/Long 40.815 N/8.425 W.

Martin's Reach WMA. Area: 89 acres. Located 0.5 mile south and 3 miles west of Wood River I-80 exit 300.
Lat/Long 40.736 N/98.641 W.

Mormon Island Crane Meadows. Area: 2,500 acres. Located 1 mile south of I-80 exit 312 on U.S. Highway 281 then west on Elm Island Road.
Lat/Long 40.492 N/98.223 W.

Mormon Island SRA. Area: 152 acres. Located 0.25 mile north of I-80 exit 312 at Grand Island. Lat/Long 40.825 N/98.371 W.

Nebraska Nature and Visitor Center. Area: 30 acres. Located at south end of I-80 Exit 305 near Alda.
Lat/Long 40.474 N/98.293 W.

HAMILTON COUNTY

Deep Well WMA. Area: 240 acres. Located 3.5 miles south of Phillips.
Lat/Long 40.845 N/98.221 W.

Gadwall WMA. Area: 90 acres. Located 1 mile west and 4.5 miles north of Aurora.
Lat/Long 40.940 N/98.037 W.

Nelson WPA. Area 143 acres. Located 3 miles north of Stockham.
Lat/Long 40.759 N/97.937 W.

Pintail WMA. Area: 500 acres. Located 2.5 miles south and 2 miles east of the Aurora I-80 exit 332.
Lat/Long 40.785 N/97.954 W.

Springer WPA. Area: 397 acres. Located 6 miles west and 1 mile south of Aurora.
Lat/Long 40.849 N/98.128 W.

Troester Basin WPA. Area: 150 acres. Located 4.5 miles north and 0.5 mile east of Stockham. Lat/Long 40.797 N/97.924 W

KEARNEY COUNTY

Bluestem WPA. Area: 76 acres. Located 3 miles south and 5 miles east of Axtell.
Lat/Long 40.441 N/99.058 W.

Clark WPA. Area: 450 acres. Located 3 miles north of Hildreth.
Lat/Long 40.378 N/99.053 W.

Fort Kearney SRA. Area: 163 acres. Located 4 miles south and 5 miles east of Kearney.
Lat/Long 40.655 N/98.996 W.

Frerichs WPA. Area: 43 acres. Located 2 miles east and 0.5 mile north of Wilcox.
Lat/Long 40.372 N/99.125 W.

Gleason WPA. Area: 570 acres. Located 4 miles south and 4 miles west of Minden.
Lat/Long 40.435 N/99.025 W.

Jensen WPA. Area: 465 acres. Located 6 miles north of Campbell.
Lat/Long 40.400 N/97.744 W.

Killdeer Basin WPA. Area: 36 acres. Located 3 miles west and 3.6 miles north of Hildreth.
Lat/Long 40.389 N/99.104 W

Lindau WPA. Area: 152 acres. Located 5 miles north and 0.5 mile east of Hildreth.
Lat/Long 40.402 N/99.036 W.

Prairie Dog WPA. Area: 471 acres. Located 5.5 miles south of Axtel, or 2 miles east and 2 miles north of Wilcox.
Lat/Long 40.402 N/99.130 W.

Youngson WPA. Area: 113 acres. Located 6 miles south and 5 miles east of Norman.
Lat/Long 40.396 N/98.785 W.

LINCOLN COUNTY

Buffalo Bill's Ranch State Historic Park and SRA. Area (Park and SRA): 233 acres. Located 2 miles west and 1 mile north of the city of North Platte (intersection of U.S. Highway 83 and U.S. Highway 30).
Lat/Long 41.162 N/100.795 W.

Chester Island WMA. Area: 69 acres. Located 1 mile south and 2 miles west of I-80 exit 199.
Lat/Long 40.991 N/100.391 W.

Lake Malony SRA. Area: *ca.* 1,100 acres. Located 6 miles south of North Platte on U.S. Highway 83.
Lat/Long 41.048 N/100.800 W.

Muskrat Run WMA. Area: 224 acres. Located 6 miles east and 1 mile north of Hershey.
Lat/Long 41.196 N/100.894 W.

North River WMA. Area: 681 acres. Located 3 miles north of Hershey.
Lat/Long 41.202 N/100.983 W.

Platte WMA. Area: 242 acres. Located 6 miles east of Maxwell.
Lat/Long 41.099 N/100.654 W.

Sutherland Reservoir SRA. Area: 3,057 acres. Located 2 miles south of Sutherland on State Highway 25.
Lat/Long 41.106 N/101.131 W.

MERRICK COUNTY

Bader Memorial Park and Natural Area. Area: 200 acres. Located 3 miles south of Chapman.
Lat/Long 41.006 N/98.086 W.

PHELPS COUNTY

Atlanta WPA. Area: 1,112 acres. Located 0.5 mile north of Atlanta, or 6 miles west and 3 miles south of Holdrege.
Lat/Long 40.382 N/99.478 W.

Cottonwood WPA. Area: 550 acres. Located 1 mile north and 1.5 mile east of Bertrand.
Lat/Long 40.547 N/99.584 W.

Funk Lagoon WPA. Area: 1,989 acres. Located 1 mile east and 3 miles north of Funk.
Lat/Long 40.498 N/99.227 W.

High Basin WMA. Area: 118 acres. Located 2 miles north of Bertrand.
Lat/Long 40.565 N/99.640 W.

Johnson Lagoon WPA. Area: 578 acres. Located 7 miles north and 2.5 miles east of Holdrege.
Lat/Long 40.556 N/99.326 W.

Jones Marsh WPA. Area: 166 acres. Located 3 miles west and 3 miles south of Holdrege.
Lat/Long 40.390 N/99.432 W.

Linder WPA. Area: 160 acres, located 1 mile north and 5 miles east of Bertrand.
Lat/Long 40.541 N/99.534 W.

Sacramento-Wilcox WMA. Area (including West Sacramento, South Sacramento and Southeast Sacramento): 3,023 acres. Located about 2.5 miles west of Wilcox.
Lat/Long 40.372 N/99.241 W.

West Sacramento WMA. Area: 388 acres. Located 4 miles south and 4 miles west of Holdrege.
Lat/Long 40.364 N/99.313 W.

Literature Cited

American Ornithologists' Union (AOU). 1998. *The A.O.U. Checklist of North American Birds.* 7th ed. American Ornithologists' Union, Washington, D.C. (Supplements: *Auk*: 117: 847–856; 119: 923–932; 120: 923–931; 121:985–995; 122: 1026–1031; 123: 926–936; 124:1109–1115; 125:758–768; 126:705–714, 127: 726–744).

Andrews, R. and R. Righter. 1992. *Colorado birds.* Denver Museum of Natural History, Denver, CO.

Banks, R. C. 1977. The decline and fall of the Eskimo Curlew, or why did it go extaille. *American Birds* 35(2):127–134.

Bennett, E. V. 1968. 1967 Nebraska nesting survey. *Nebraska Bird Review* 36: 35–42.

Bennett. E. V. 1973. 1972 Nebraska nesting survey. *Nebraska Bird Review* 41: 3–9.

Bennett, E. V. 1974. 1973 Nebraska nesting survey. *Nebraska Bird Review* 42: 3–10.

Bennett, E. V. 1975. 1974 Nebraska nesting survey. *Nebraska Bird Review* 43: 13–19.

Bent, A. C. 1946. *Life histories of North American jays, crows, and titmice.* Bulletin of the United States National Museum 191. Dover Publications, New York, NY.

Bishop, A., J. Liske-Clark, and R. Grosse. 2009. *Nebraska land cover development.* Rainwater Basin Joint Venture and Great Plains GIS Partnership. Grand Island, NE.

Bleed, A. and C. Flowerday (eds.) 1989. *An Atlas of the Sand Hills.* Resource Atlas No. 5. Conservation and Survey Division, Institute of Agriculture and Natural Resources, University of Nebraska, Lincoln, NE.

Bliese, J. C. W. 1975. A Monk Parakeet in the Kearney area. *Nebraska Bird Review* 43: 42.

Branson, F. A. 1952. Native pastures of the dissected loess plains of central Nebraska. Ph.D. dissertation, University of Nebraska, Lincoln, NE.

Bray, T. E., B. K. Padelford and W.R. Silcock. 1986. *The birds of Nebraska: a critically evaluated list.* Published by the authors. Bellevue, NE.

Brogie, M. A. 1998. 1997 (Ninth) report of the NOU Records Committee. *Nebraska Bird Review* 66: 147–159.

Brooking, A. M. 1942. The vanishing birdlife of Nebraska. *Nebraska Bird Review* 10: 43–47.

Brooking, A. M. 1944. Nesting of the White-necked Raven in Kearney County. *Nebraska Bird Review* 12: 40.

Brown, E. 1947. Nests of the White-necked Raven in Kearney County. *Nebraska Bird Review* 12: 40.

Brown, S., C. Hickey, B. Harrington, and R. Gill (eds.) 2001. *The U.S. Shorebird Conservation Plan,* 2nd edition, Manomet Center for Conservation Sciences, Manomet, MA. http://www.nwtf.org/NAWTMP/downloads/Literature/US_Shorebird_Conservation_Plan.pdf

Brown, C. R. and M. B. Brown. 1996. *Coloniality in the Cliff Swallow.* University of Chicago Press, Chicago, IL.

Brown, M. B., S. Dinsmore, and C. R. Brown. 2012. *Birds of Southwestern Nebraska.* Lincoln, NE: Conservation and Survey Division, Institute of Agriculture and Natural Resources, University of Nebraska, Lincoln, NE.

Bruner, L. 1896. *A list of Nebraska birds, together with notes on their abundance, migrations, breeding, food-habits, etc.* Nebraska State Horticultural Society 27th annual report. Lincoln, NE.

Bruner, L., R. H. Wolcott and M. H. Swenk. 1904. *A preliminary review of the birds of Nebraska, with synopses.* Klopp and Bartlett, Omaha, NE.

Cink, C. L. 1973. Louisiana Heron in Clay County. *Nebraska Bird Review* 41:14–15.

Colt, C. J. 1995. Breeding bird use of riparian forests along the central Platte River: A spatial analysis. M.S. thesis, University of Nebraska, Lincoln, NE.

Currier, P. J., 1981. The floodplain vegetation of the Platte River; forest development and seeding establishment, Ph.D., dissertation, Iowa State University, Ames, IA.

Davis, C. A. 2001. Abundance and habitat association of birds wintering in the Platte River Valley, Nebraska. *Great Plains Research* 11: 233–248.

Davis, C. A. 2005. Breeding bird communities in riparian forests along the central Platte River, Nebraska. *Great Plains Research* 15: 199–211.

Dinsmore, S. J. and W. R. Silcock. 1993. First record of a Ross' Gull for Nebraska. *Nebraska Bird Review* 61: 88–89.

Donaldson, G., C. Hyslop, G. Morrison, L. Dickson, and I. Davidson. 2000. *Canadian Shorebird Conservation Plan.* Canadian Wildlife Service, Environment Canada, Ottawa, ON. http://fresc.usgs.gov/products/blackoyster-catcher/conservation_plans/Canadian_Shorebird_Conserv_Plan.pdf

Ducey, J. E. 1988. *Nebraska Birds: Breeding Status and Distribution.* Simmons-Boardman Books, Omaha, NE.

Ducey, J. E. 2000. *Birds of the Untamed West: the History of Birdlife in Nebraska, 1750–1875,* Making History Books, Omaha, NE.

ECONorthwest. 2006. *Natural-Resource Amenities and Nebraska's Economy: Current Connections, Challenges and Possibilities.* Eugene, OR: ECONorthwest. 120 pp.

Eubanks, T., L., Jr., R. B. Ditotn, and J. R. Stoll, 1998. *Platte River Nature Recreation Study.* Fermata, Inc., Austin, TX.

Evans, R. D., and C. W. Wolfe. 1967. Waterfowl production in the Rainwater Basin area of Nebraska. *Journal of Wildlife Management* 33:788–794.

Faanes, C., and G. R. Lingle. 1995. *Breeding Birds of the Platte Valley of Nebraska.* Northern Prairie Wildlife Research Center, Jamestown, ND. http://www.npwrc.usgs.gov/resources/distr/birds/platte/platte

Farrar, J. 1982. The Rainwater Basin: Nebraska's vanishing wetlands. *Nebraskaland* 60(3): 18–41.

Farrar, J. (ed.) 1985. Birds of Nebraska. *Nebraskaland* 63(1):1–146.

Farrar, J. 1996. Nebraska's Rainwater Basin. *Nebraskaland* 72(4):18–35.

Farrar, J. 1997. Loess Hills. *Nebraskaland* 75(7):14–25.

Farrar, J. 2004. Birding Nebraska. *Nebraskaland* 82(1):1–178.

Fermata. Inc. 1996. Platte River Nature Recreation Study: Executive Summary, http://fermatainc.com/basic/eco_nebplatte.html

Freeman. D. M. 2010. *Implementing the Endangered Species Act on the Platte Basin Commons.* University Press of Colorado, Boulder, CO.

Freeman, P. 1989. "Mammals," in *An Atlas of the Sand Hills* (A. Bleed and C. Flowerday, eds.). pp. 181–188, Resource Atlas No. 5. Conservation and Survey Division, Institute of Agriculture and Natural Resources, University of Nebraska, Lincoln, NE.

Geluso, K., and M. J. Harner. 2011. Reexamination of herpetofauna on Mormon Island, Hall County, Nebraska, with notes on their natural history. in Responses of Herpetofauna to Grazing and Fire in Wet, Tallgrass Prairies along the Platte River (M. Harner and K. Geluso, eds.). pp. 20–56, Report to the Nebraska Game and Parks Commission, Lincoln, NE.

Grenon, A. G. 1990. 1990 (Third) Report of the NOU Records Committee. *Nebraska Bird Review* 58: 90–97.

Gubanyi. 1996. 1994 (Sixth) Report of the NOU Records Committee. *Nebraska Bird Review* 64: 30–35.

Haecker, F. W., R. A. Moser and J. B. Swenk. 1945. Checklist of the birds of Nebraska. *Nebraska Bird Review* 13: 1–40.

Haig, S. M. 1992. "Piping Plover," in *The Birds of North America*, No. 2 (A. Poole, P. Stettenheim and F. Gill, eds.). The Academy of Natural Sciences, Philadelphia, PA and American Ornithologists' Union, Washington, D. C.

Haig, S. M., W. Harrison, R. Lock, L. Pfannmuller, Pike, E., M. Ryan, and J. Sidle. 1988. Recovery plan for Piping Plovers (*Charadrius melodus*) of the Great Lakes and Northern Great Plains. United States Fish and Wildlife Service. Twin Cities, MN.

Harding, R. G. 1986. Waterfowl nesting preferences and productivity in the Rainwater Basin, Nebraska. M.S. thesis, Kearney State College, Kearney, NE.

Hopkins, H. H. 1949. Ecology of the native vegetation of the Loess Hills in central Nebraska. Ph.D. dissertation, University of Nebraska, Lincoln, NE.

Johnsgard, P. A. 1974. *Song of the North Wind: A Story of the Snow Goose.* New York: Doubleday.

Johnsgard, P. A. 1995. *This Fragile Land: A Natural History of the Nebraska Sandhills.* University of Nebraska Press, Lincoln, NE.

Johnsgard, P. A. 1997. *The Avian Brood Parasites: Deception at the Nest.* Oxford University Press, New York, NY.

Johnsgard, P. A. 2001. *The Nature of Nebraska.* University of Nebraska Press, Lincoln, NE.

Johnsgard, P. A. 2005. *The Platte: Channels in Time.* 2nd edition. University of Nebraska Press, Lincoln, NE.

Johnsgard, P. A. 2007. *The Birds of Nebraska.* Revised edition. University of Nebraska Digital Commons, Lincoln, NE. http://digitalcommons.unl.edu/biosciornithology/38

Johnsgard, P. A. 2008. *A Guide to the Natural History of the Central Platte Valley of Nebraska.* University of Nebraska Digital Commons, Lincoln, NE. http://digitalcommons.unl.edu/biosciornithology/40

Johnsgard, P. A. 2009. *Birds of the Great Plains: Breeding Species and their Distribution.* Revised edition. University of Nebraska Digital Commons. Lincoln, NE. http://digitalcommons.unl.edu/bioscibirdsgreatplains/1/

Johnsgard, P. A. 2011. *A Nebraska Bird-finding Guide.* Zea E-Books, University of Nebraska Libraries, Lincoln, NE. http://digitalcommons.unl.edu/zeabook/5/

Johnsgard, P. A. 2012. *Wetland Birds of the Central Plains: South Dakota, Nebraska and Kansas.* University of Nebraska Digital Commons and Zea Press, Lincoln, NE. http://digitalcommons.unl.edu/zeabook/8/

Johnsgard, P. A. 2012. *Nebraska's Wetlands: Their Wildlife and Ecology.* Conservation and Survey Division, Institute of Agriculture and Natural Resources, University of Nebraska, Lincoln, NE.

Johnsgard, P. A. and K. Gil. 2010. The whooping cranes: Survivors against all odds. *Prairie Fire,* Sept., 2010 http://www.prairiefirenewspaper.com/2010/9/the-whooping-cranes-survivors-against-all-odds

Johnson, W. C. 1994. Woodland expansion in the Platte River, Nebraska: patterns and causes. *Ecological Monographs* 64: 45–84.

Jones, A. H. 1939. The Williamson's Sapsucker at Hastings, Adams County. *Nebraska Bird Review* 7: 27–28.

Jones, S. M., R. E. Ballinger and J. W. Niefeldt. 1981. Herpetofauna of Mormon Island Preserve, Hall County, Nebraska. *Prairie Naturalist* 13:33–41.

Jorgensen, J. G. 1994. Ruff with godwits. *Nebraska Bird Review* 62: 98–99.

Jorgensen, J. G. 1996. A review of the status of *Limnodromus griseus*, the Short-billed Dowitcher, in Nebraska. *Nebraska Bird Review* 64: 74–78.

Jorgensen, J. G. 2002. 2000 (12th) Report of the NOU Records Committee. *Nebraska Bird Review* 70: 84-90.

Jorgensen. J. G. 2004. *An Overview of the Shorebird Migration in the Eastern Rainwater Basin, Nebraska.* Nebraska Ornithologists' Union Occasional Paper No. 8, Lincoln, NE.

Jorgensen. J. G. 2012. *Birds of the Rainwater Basin, Nebraska.* Nebraska Game and Parks Commission, Lincoln, NE. http://outdoornebraska.ne.gov/wildlife/programs/nongame/NGBirds/pdf/Birds%20of%20the%20Rainwater%20Basin%20Version%201.0%20(May%202012).pdf

Jorgensen, J. G. and W. R. Silcock. 1998. Nebraska's first Curlew Sandpiper (*Calidris ferruginea*). *Nebraska Bird Review* 66: 3.

Kaul, R. B., D. Sutherland, and S. B. Rolfsmeier. 2006. *The Flora of Nebraska.* Conservation and Survey Division, Institute of Agriculture and Natural Resources, University of Nebraska, Lincoln, NE.

Kim, D. H. 2005. First Nebraska nest record for Henslow's Sparrow. *Prairie Naturalist* 37: 171–173.

Kim, D. H., W. E. Newton, G. R. Lingle, and F. Chavez-Ramirez. 2008. Influence of grazing and available moisture on breeding densities of grassland birds in the central Platte Valley, Nebraska. *Wilson Journal of Ornithology* 120: 820–829.

Kimball, B. 1984. Chuck-will's-widow. *Nebraska Bird Review* 52: 24.

Kuzila, M. S. 1994. Inherited morphologies of two large basins in Clay County, Nebraska. *Great Plains Research* 4:51–63.

Labedz, T. 1989. "Birds," in *An Atlas of the Sand Hills* (A. Bleed and C. Flowerday, eds.). pp. 161–180, Resource Atlas No. 5. Conservation and Survey Division, Institute of Agriculture and Natural Resources, University of Nebraska, Lincoln, NE.

LaGrange, T. 2005. *Guide to Nebraska's wetlands and their conservation needs,* 2nd edition. Nebraska Game and Parks Commission, Lincoln, NE.

Lingle, G. R. l992. History and economic impact of crane-watching in central Nebraska. *Proceedings North American Crane Workshop* 6:33–37.

Lingle, G. R. 1994. *Birding Crane River: Nebraska's Platte.* Harrier Publishing, Grand Island, NE.

Loope, D. B., and J. B. Swinehart. 2000. Thinking like a dune field: Geologic history in the Nebraska Sand Hills. *Great Plains Research* 10:5–35.

Maunder, V. 1996. Roseate Spoonbill. *Nebraska Bird Review* 34: 77.

Mollhoff, W. J. 2001. *The Nebraska Breeding Bird Atlas 1984–1989.* Nebraska Ornithologists' Union Occasional Papers No. 7 and Nebraska Game and Parks Commission Nebraska Technical Series No. 20, Lincoln, NE.

Moulton, G. 2005. *The Journals of the Lewis and Clark Expedition.* University of Nebraska Press-University of Nebraska-Lincoln Libraries, Lincoln, NE. http://lewisandclarkjournals.unl.edu

Nagel, H. G. 1981. Vegetation ecology of Crane Meadows. Final report. The Nature Conservancy and Platte River Whooping Crane Habitat Maintenance Trust, Alda, NE.

Nagel, H. G., and O. A. Kolstad. 1987. Comparison of plant species composition of Mormon Island Crane Meadows and Lillian Annette Rowe Sanctuary in central Nebraska. *Transactions of the Nebraska Academy of Science* 15:37–48.

National Research Council of the National Academies. 2005. *Endangered and Threatened Species of the Platte River.* National Academies Press, Washington, D.C.

Novacek, J. M. 1989. The water and wetland resources of the Nebraska Sandhills. in *Northern Prairie Wetlands* (A. van der Valk, ed.). pp. 340–384, Iowa State University Press, Ames, IA.

Oberholser, H. C, and W. L. McAtee. 1920. Waterfowl and their food plants in the Sandhills region of Nebraska. *United States Department of Agriculture Bulletin* 794:1–79.

Paine, E. 1988. An Inca Dove in Nebraska, in winter. *Nebraska Bird Review* 56: 3.

Rapp, W. F., Jr., J. L. C. Rapp, H. E. Baumgarten, and R. A. Moser. 1958. Revised checklist of Nebraska birds. Occasional Papers 5. Nebraska Ornithologists' Union, Crete, NE.

Rising, J. D. 1968. The Great Plains hybrid zones. *Current Ornithology* 1:131–157.

Rising, J. D. 1970. Morphological variation and evolution in some North American orioles. *Systematic Zoology* 19: 315–351.

Robbins, M. B. and D. A. Easterla. 1992. *Birds of Missouri, their distribution and abundance.* University of Missouri Press, Columbia, MO.

Robertson, K. 1977. Jaeger. *Nebraska Bird Review* 45: 15.

Rolfsmeier, S. B. and G. Steinauer. 2010. *Terrestrial Ecological System and Natural Communities of Nebraska.* Nebraska Game and Parks Commission, Lincoln, NE.

Rothenberger, S. 1998. "Vegetation of the Loess Hills," in *The Loess Hills Prairies of Central Nebraska* (Nagel and Plambeck, eds.). pp. 63–73, University of Nebraska, Kearney, NE.

Rothenberger, S. J., and C. J. Bicak. 1993. Plants. in *The Platte River: An Atlas of the Big Bend Region.* pp. 18–28. University of Nebraska, Kearney, NE.

Scharf, W. C. 2007. Woodland bird use of in-channel islands in the central Platte River, Nebraska. *Prairie Naturalist* 39:15–28.

Schneider, R., K. Stoner, G. Steinhauer, M. Panella, and M. Humpert. 2011. *The Nebraska Natural Legacy Project State Wildlife Action Plan,* 2nd edition, Nebraska Game and Parks Commission. Lincoln, NE.

Sharpe, R. W., R. Silcock, and J. G. Jorgensen. 2001. *The Birds of Nebraska, their distribution and temporal occurrence.* University of Nebraska Press, Lincoln, NE.

Shickley, G. M. 1965. Calliope Hummingbird in Nebraska. *Auk* 82: 650.

Shickley, G. M. 1968. Additions to the Lincoln County checklist. *Nebraska Bird Review* 36:54–57.

Short, L. L., Jr. 1961. Notes on bird distribution in the central Plains. *Nebraska Bird Review* 29: 2–22.

Sibley, C. G. and L. L. Short, Jr.. 1959. Hybridization in the buntings (*Passerina*) of the Great Plains. *Auk* 76: 443–463.

Sibley, C. G. and D. A. West. 1959. Hybridization in the Rufous-sided Towhees of the Great Plains. *Auk* 76: 326–328.

Sidle, J. G. and W. F. Harrison. 1990. Recovery plan for the Interior population of the Least Tern (*Sterna antillarum*). United States Fish and Wildlife Service, Twin Cities, MN.

Sidle, J. G., E. D. Miller and P. Currier. 1989. Changing habitats on the Platte. *Prairie Naturalist* 21:91–104.

Silcock, W. R. 2001. Summer field report. *Nebraska Bird Review* 69: 106–132.

Skagen, S. K., P. B. Sharpe, R. G. Waltermire, and M.B. Dillon. 1999. Biogeographical profiles of shorebird migration in midcontinental North America. Biological Science Report USGS/BRD/BSR-2000-0003, Denver, CO.

Stoppkotte, G. W. 1975. A Groove-billed Ani seen again in Nebraska. *Nebraska Bird Review* 44: 79–80.

Swenk, M. H. 1915a. The Eskimo Curlew and its disappearance. Annual Report to the Smithsonian Institution for 1915. Washington, D. C.

Swenk, M. H. 1915b. The birds and mammals of Nebraska. Nebraska Blue Book, State of Nebraska, Lincoln, NE.

Swenk, M. H. 1937. A study of the distribution and migration of the Great Horned Owls in the Missouri Valley region. *Nebraska Bird Review* 5: 79–105.

Swenk, M. H. and R. W. Dawson. 1921. Notes on the distribution and migration of Nebraska birds. I. Tyrant Flycatchers (Tyrannidae). *Wilson Bulletin* 33: 132–141.

Thompson, B. C., J. A. Jackson, J. Burger, L. A. Hill, E. M. Kirsch, and J. L. Atwood. 1997. "Least Tern," in *Birds of North America* No. 290 (A. Poole, P. Stettenheim and F. Gill, eds.). The Academy of Natural Sciences, Philadelphia, PA and American Ornithologists' Union, Washington, D. C.

Tout, W. 1902. Ten years without a gun. *Proceedings of the NOU* 3: 42–45.

Tout, W. 1947. *Lincoln County Birds*. Published by the author. North Platte, NE.

United States Department of the Interior–United States Fish and Wildlife Service. 2006. Platte River Recovery Program: Final Environmental Impact Statement. Washington, D. C.

United States Fish and Wildlife Service (U.S.F.W.S). 2005. Rainwater Basin Wetland Management District Bird List. United States Department of the Interior, Washington, D.C.

Viehmeyer, G. 1961. Calliope Hummingbird at North Platte. *Nebraska Bird Review* 29: 39–40.

Vrtiska, M., and S. Sullivan. 2009. Abundance and distribution of Lesser Snow and Ross's Geese in the Rainwater Basin and Central Platte Valley of Nebraska. *Great Plains Research* 19:147–155.

Weakley, H. E. 1936. The Palmer Curve-billed Thrasher at North Platte, Lincoln County. *Nebraska Bird Review* 4: 54.

Wehrman, K. C. 1961. A study of the transition zone between the Loess Hills and the Sand Hills in central Nebraska. M.S. thesis, University of Nebraska, Lincoln, NE.

Wiens, J, A. 1973. Pattern and process in grassland bird communities. *Ecological Monographs* 8:1–93.

Wiens, J, A., and M. J. Dyer. 1975. Rangeland avifaunas: Their composition, energetics, and role in the ecosystem. *Proceedings of a Symposium on the Management of Forest and Range Habitats for Nongame Birds.* USDA Forest Service General Technical Report. WO-1:146–182.

Wiles, M. R. and B. S. Goldowitz. 2005. Macroinvertebrate communities in central Platte wetlands: patterns across a hydrologic gradient. *Wetlands* 25:462–472.

Wright, R. 1983. Olivaceous Cormorant. *Nebraska Bird Review* 51: 18.

Check-List of Species

Order Anseriformes

Family Anatidae: Geese, Swans, and Ducks

- ☐ Black-bellied Whistling-Duck, *Dendrocyna autumnalis*
- ☐ Taiga Bean-Goose, *Anser fabalis*
- ☐ [Pink-footed Goose, *Anser brachyrhynchus*]
- ☐ Pink-footed Goose, *Anser brachyrhynchus*
- ☐ Greater White-fronted Goose, *Anser albifrons*
- ☐ [Swan Goose, *Anser cygnoides*]
- ☐ [Emperor Goose, *Chen canagica*]
- ☐ Snow Goose, *Chen caerulescens*
- ☐ Ross's Goose, *Chen rossii*
- ☐ Brant, *Branta bernicla*
- ☐ [Barnacle Goose, *Branta leucopsis*]
- ☐ Cackling Goose, *Branta hutchinsii*
- ☐ Canada Goose, *Branta canadensis*
- ☐ Trumpeter Swan, *Cygnus buccinator*
- ☐ Tundra Swan, *Cygnus columbianus*
- ☐ [Mute Swan, *Cygnus olor*]
- ☐ [Ruddy Shelduck, *Tadorna ferriginea*]
- ☐ Wood Duck, *Aix sponsa*
- ☐ Gadwall, *Anas strepera*
- ☐ Eurasian Wigeon, *Anas penelope*
- ☐ American Wigeon, *Anas americana*
- ☐ American Black Duck, *Anas rubripes*
- ☐ Mallard, *Anas platyrhynchos*
- ☐ Blue-winged Teal, *Anas discors*
- ☐ Cinnamon Teal, *Anas cyanoptera*
- ☐ Northern Shoveler, *Anas clypeata*
- ☐ Northern Pintail, *Anas acuta*
- ☐ Green-winged Teal, *Anas crecca*

☐ Canvasback, *Aythya valisineria*
☐ Redhead, *Aythya americana*
☐ Ring-necked Duck, *Aythya collaris*
☐ Greater Scaup, *Aythya marila*
☐ Lesser Scaup, *Aythya affinis*
☐ Common Eider, *Somateria mollissima*
☐ Surf Scoter, *Melanitta perspicillata*
☐ White-winged Scoter, *Melanitta fusca*
☐ Black Scoter, *Melanitta americana*
☐ Long-tailed Duck, *Clangula hyemalis*
☐ Bufflehead, *Bucephala albeola*
☐ Common Goldeneye, *Bucephala clangula*
☐ Barrow's Goldeneye, *Bucephala islandica*
☐ Hooded Merganser, *Lophodytes cucullatus*
☐ Common Merganser, *Mergus merganser*
☐ Red-breasted Merganser, *Mergus serrator*
☐ Ruddy Duck, *Oxyura jamaicensis*

Order Galliformes

Family Odontophoridae: New World Quail

☐ Northern Bobwhite, *Colinus virginianus*

Family Phasianidae: Pheasants, Grouse and Turkeys

☐ [Coturnix Quail, *Coturnix coturnix*]
☐ [Chukar, *Alectoris chukar*]
☐ Gray Partridge, *Perdix perdix*
☐ Ring-necked Pheasant, *Phasianus colchicus*
☐ Sharp-tailed Grouse, *Tympanuchus phasianellus*
☐ Greater Prairie-Chicken, *Tympanuchus cupido*
☐ Wild Turkey, *Meleagris gallopavo*

Order Gaviiformes

Family Gaviidae: Loons

☐ Red-throated Loon, *Gavia stellata*.
☐ Common Loon, *Gavia immer*

Order Podicipediformes

Family Podicipedidae: Grebes

- ☐ Pied-billed Grebe, *Podilymbus podiceps*
- ☐ Horned Grebe, *Podiceps auritus*
- ☐ Eared Grebe, *Podiceps nigricollis*
- ☐ Western Grebe, *Aechmophorus occidentalis*
- ☐ Clark's Grebe, *Aechmophorus clarkii*

Order Phaethontiformes

Family Phaethontidae: Tropicbirds

- ☐ [White-tailed Tropicbird, *Phaethon lepturus*]

Order Suliformes

Family Phalacrocoracidae: Cormorants

- ☐ Neotropic Cormorant, *Phalacrocorax brasilianus*
- ☐ Double-crested Cormorant, *Phalacrocorax auritus*

Order Pelecaniformes

Family Pelecanidae: Pelicans

- ☐ American White Pelican, *Pelecanus erythrorhynchos*
- ☐ Brown Pelican, *Pelecanus occidentalis*

Family Ardeidae: Bitterns and Herons

- ☐ American Bittern, *Botaurus lentiginosus*
- ☐ Least Bittern, *Ixobrychus exilis*
- ☐ Great Blue Heron, *Ardea herodias*
- ☐ Great Egret, *Ardea alba*
- ☐ Snowy Egret, *Egretta thula*
- ☐ Little Blue Heron, *Egretta caerulea*
- ☐ Tricolored Heron, *Egretta tricolor*
- ☐ Reddish Egret, *Egretta rufescens*
- ☐ Cattle Egret, *Bubulcus ibis*
- ☐ Green Heron, *Butorides virescens*
- ☐ Black-crowned Night-Heron, *Nycticorax nycticorax*
- ☐ Yellow-crowned Night-Heron, *Nyctanassa violacea*

Family Threskiornithidae: Ibises and Spoonbills

- ☐ White Ibis, *Eudomimus albus*
- ☐ White-faced Ibis, *Plegadis chihi*
- ☐ Glossy Ibis, *Plegadis falcinellis*
- ☐ Roseate Spoonbill, *Platalea ajaja*

Order Accipitriformes

Family Cathartidae: New World Vultures

- ☐ [Black Vulture, *Coragyps atratus*]
- ☐ Turkey Vulture, *Cathartes aura*

Family Pandionidae: Ospreys

- ☐ Osprey, *Pandion haliaetus*

Family Accipitridae: Kites, Hawks and Eagles

- ☐ Swallow-tailed Kite, *Elanoides forficatus*
- ☐ White-tailed Kite, *Elanus leucurus*
- ☐ Mississippi Kite. *Ictinia mississippiensis*
- ☐ Bald Eagle, *Haliaeetus leucocephalus*
- ☐ Northern Harrier, *Circus cyaneus*
- ☐ Sharp-shinned Hawk, *Accipiter striatus*
- ☐ Cooper's Hawk, *Accipiter cooperii*
- ☐ Northern Goshawk, *Accipiter gentilis*
- ☐ Red-shouldered Hawk, *Buteo lineatus*
- ☐ Broad-winged Hawk, *Buteo platypterus*
- ☐ Swainson's Hawk, *Buteo swainsoni*
- ☐ Red-tailed Hawk, *Buteo jamaicensis*
- ☐ Ferruginous Hawk, *Buteo regalis*
- ☐ Rough-legged Hawk, *Buteo lagopus*
- ☐ Golden Eagle, *Aquila chrysaetos*

Order Falconiformes

Family Falconidae: Falcons

- ☐ American Kestrel, *Falco sparverius*
- ☐ Merlin, *Falco columbarius*

☐ Gyrfalcon, *Falco rusticus*
☐ Peregrine Falcon, *Falco peregrinus*
☐ Prairie Falcon, *Falco mexicanus*

Order Gruiformes

Family Rallidae: Rails, Gallinules and Coots

☐ Yellow Rail, *Coturnicops noveboracensis*
☐ Black Rail, *Laterallus jamaicensis*
☐ King Rail, *Rallus elegans*
☐ Virginia Rail, *Rallus limicola*
☐ Sora, *Porzana carolina*
☐ Common Moorhen, *Gallinula galeata*
☐ American Coot, *Fulica americana*

Family Gruidae: Cranes

☐ Sandhill Crane, *Grus canadensis*
☐ Common Crane, *Grus grus*
☐ Whooping Crane, *Grus americana*
☐ Hooded Crane, *Grus monacha*

Order Charadriiformes

Family Charadriidae: Plovers

☐ Black-bellied Plover, *Pluvialis squatarola*
☐ American Golden-Plover, *Pluvialis dominica*
☐ Snowy Plover, *Charadrius nivosus*
☐ Mountain Plover, *Charadrius montana*
☐ Semipalmated Plover, *Charadrius semipalmatus*
☐ Piping Plover, *Charadrius melodus*
☐ Killdeer, *Charadrius vociferus*

Family Recurvirostridae: Stilts and Avocets

☐ Black-necked Stilt, *Himantopus mexicanus*
☐ American Avocet, *Recurvirostra americana*

Family Scolopacidae: Sandpipers and Phalaropes

☐ Spotted Sandpiper, *Tringa macularia*
☐ Solitary Sandpiper, *Tringa solitaria*
☐ Greater Yellowlegs, *Tringa melanoleuca*
☐ Willet, *Tringa semipalmata*
☐ Lesser Yellowlegs, *Tringa flavipes*
☐ Upland Sandpiper, *Bartramia longicauda*
☐ Eskimo Curlew, *Numenius borealis*
☐ Whimbrel, *Numenius phaeopus*
☐ Long-billed Curlew, *Numenius americanus*
☐ Hudsonian Godwit, *Limosa haemastica*
☐ Marbled Godwit, *Limosa fedoa*
☐ Ruddy Turnstone, *Arenaria interpres*
☐ Red Knot, *Calidris canutus*
☐ Sanderling, *Calidris alba*
☐ Semipalmated Sandpiper, *Calidris pusilla*
☐ Western Sandpiper, *Calidris mauri*
☐ Least Sandpiper, *Calidris minutilla*
☐ White-rumped Sandpiper, *Calidris fuscicollis*
☐ Baird's Sandpiper, *Calidris bairdii*
☐ Pectoral Sandpiper, *Calidris melanotos*
☐ Sharp-tailed Sandpiper, *Calidris acuminata*
☐ Dunlin, *Calidris alpina*
☐ Curlew Sandpiper, *Calidris ferruginea*
☐ Stilt Sandpiper, *Calidris himantopus*
☐ Buff-breasted Sandpiper, *Tryngites subruficollis*
☐ Ruff, *Philomachus pugnax*
☐ Short-billed Dowitcher, *Limnodromus griseus*
☐ Long-billed Dowitcher, *Limnodromus scolopaceus*
☐ Wilson's Snipe, *Gallinago delicata*
☐ American Woodcock, *Scolopax minor*
☐ Wilson's Phalarope, *Phalaropus tricolor*
☐ Red-necked Phalarope, *Phalaropus lobatus*
☐ Red Phalarope, *Phalaropus fulicaria*

Family Laridae: Gulls and Terns

- ☐ [Jaeger species, *Stercorarius* spp]
- ☐ Pomarine Jaeger, *Stercorarius pomarinus*
- ☐ Black-legged Kittiwake, *Rissa tridactyla*
- ☐ Sabine's Gull, *Xema sabini*
- ☐ Little Gull, *Hydrocoleus minutus*
- ☐ Ross's Gull, *Rhodostethia rosea*
- ☐ Laughing Gull. *Leucophaeus atricilla*
- ☐ Franklin's Gull, *Leucophaeus pipixcan*
- ☐ Mew Gull, *Larus canus*
- ☐ Ring-billed Gull, *Larus delawarensis*
- ☐ California Gull, *Larus californicus*
- ☐ Herring Gull, *Larus argentatus*
- ☐ Thayer's Gull, *Larus thayeri*
- ☐ Lesser Black-backed Gull, *Larus fuscus*
- ☐ Glaucous Gull, *Larus hyperboreus*
- ☐ Great Black-backed Gull, *Larus marinus*
- ☐ Least Tern, *Sterna antillarum*
- ☐ Caspian Tern, *Sterna caspia*
- ☐ Black Tern, *Chlidonias niger*
- ☐ Common Tern, *Sterna hirundo*
- ☐ Forster's Tern, *Sterna forsteri*

Order Columbiformes

Family Columbidae: Pigeons and Doves

- ☐ Rock Pigeon, *Columba livia*
- ☐ Eurasian Collared-Dove, *Streptopelia decaocto*
- ☐ White-winged Dove, *Zenaida asiatica*
- ☐ Mourning Dove, *Zenaida macroura*
- ☐ Inca Dove, *Columbina inca*
- ☐ Common Ground Dove, *Columbina passerina*

Order Psittaciformes

Family Psittacidae: Parrots

☐ [Monk Parakeet, *Myopsitta monachus*]
☐ [Budgerigar, *Melopsittacus undulatus*]

Order Cuculiformes

Family Cuculidae: Cuckoos

☐ Yellow-billed Cuckoo, *Coccyzus americanus*
☐ Black-billed Cuckoo, *Coccyzus erythropthalmus*
☐ Groove-billed Ani, *Crotophaga sulcirostris*

Order Strigiformes

Family Tytonidae: Barn Owls

☐ Barn Owl, *Tyto alba*

Family Strigidae: Typical Owls

☐ Flammulated Owl, *Otus flammeolus*
☐ Eastern Screech-Owl, *Megascops asio*
☐ Great Horned Owl, *Bubo virginianus*
☐ Snowy Owl, *Bubo scandiaca*
☐ Burrowing Owl, *Athene cunicularia*
☐ Barred Owl, *Strix varia*
☐ Great Gray Owl, *Strix nebulosa*
☐ Long-eared Owl, *Asio otus*
☐ Short-eared Owl, *Asio flammeus*
☐ Northern Saw-whet Owl, *Aegolius acadicus*
☐ Boreal Owl, *Aegolius funearius*

Order Caprimulgiformes

Family Caprimulgidae: Goatsuckers

☐ Common Nighthawk, *Chordeiles minor*
☐ Common Poorwill, *Phalaenoptilus nuttallii*
☐ Chuck-will's-widow, *Caprimulgus carolinensis*
☐ Whip-poor-will. *Caprimulgus vociferus*

Order Apodiformes

Family Apodidae: Swifts

☐ Chimney Swift, *Chaetura pelagica*

Family Trochilidae: Hummingbirds

☐ Ruby-throated Hummingbird, *Archilochus colubris*
☐ Costa's Hummingbird, *Calypte costae*
☐ Calliope Hummingbird, *Stellula calliope*
☐ Broad-tailed Hummingbird, *Selasphorus platycercus*
☐ Rufous Hummingbird, *Selasphorus rufus*

Order Coraciiformes

Family Alcedinidae: Kingfishers

☐ Belted Kingfisher, *Ceryle alcyon*

Order Piciformes

Family Picidae: Woodpeckers

☐ Lewis's Woodpecker, *Melanerpes lewis*
☐ Red-headed Woodpecker, *Melanerpes erythrocephalus*
☐ Red-bellied Woodpecker, *Melanerpes carolinus*
☐ Williamson's Sapsucker, *Sphyrapicus thyroideus*
☐ Yellow-bellied Sapsucker, *Sphyrapicus varius*
☐ Downy Woodpecker, *Picoides pubescens*
☐ Hairy Woodpecker, *Picoides villosus*
☐ Northern Flicker, *Colaptes auratus*

Order Passeriformes

Family Tyrannidae: American Flycatchers

☐ Olive-sided Flycatcher, *Contopus cooperi*
☐ Western Wood-Pewee, *Contopus sordidulus*
☐ Eastern Wood-Pewee, *Contopus virens*
☐ Yellow-bellied Flycatcher, *Empidonax flaviventris*
☐ [Acadian Flycatcher], *Empidonax virescens*

- ☐ Alder Flycatcher, *Empidonax alnorum*
- ☐ Willow Flycatcher, *Empidonax traillii*
- ☐ Least Flycatcher, *Empidonax minimus*
- ☐ Hammond's Flycatcher, *Empidonax hammondii*
- ☐ Eastern Phoebe, *Sayornis phoebe*
- ☐ Say's Phoebe, *Sayornis saya*
- ☐ Vermillion Flycatcher, *Pyrocephalus rubinus*
- ☐ Great Crested Flycatcher, *Myiarchus crinitus*
- ☐ Cassin's Kingbird, *Tyrannis vociferans*
- ☐ Western Kingbird, *Tyrannus verticalis*
- ☐ Eastern Kingbird, *Tyrannus tyrannus*
- ☐ Scissor-tailed Flycatcher, *Tyrannus forficatus*

Family Laniidae: Shrikes

- ☐ Loggerhead Shrike, *Lanius ludovicianus*
- ☐ Northern Shrike, *Lanius excubitor*

Family Vireonidae: Vireos

- ☐ Bell's Vireo, *Vireo bellii*
- ☐ Yellow-throated Vireo, *Vireo flavifrons*
- ☐ Cassin's Vireo. *Vireo cassinii*
- ☐ Blue-headed Vireo, *Vireo solitarius*
- ☐ Warbling Vireo, *Vireo gilvus*
- ☐ Philadelphia Vireo, *Vireo philadelphicus*
- ☐ Red-eyed Vireo, *Vireo olivaceus*
- ☐ White-eyed Vireo, *Vireo griseus*

Family Corvidae: Jays, Magpies, and Crows

- ☐ Pinyon Jay, *Gymnorhinus cyanocephalus*
- ☐ Steller's Jay, *Cyanocitta stelleri*
- ☐ Blue Jay, *Cyanocitta cristata*
- ☐ Western Scrub-Jay, *Aphelocoma californica*
- ☐ Clark's Nutcracker, *Nucifraga columbiana*
- ☐ Black-billed Magpie, *Pica hudsonia*

☐ American Crow, *Corvus brachyrhynchos*
☐ Chihuahuan Raven, *Corvus cryptoleucus*
☐ Common Raven, *Corvus corax*

Family Alaudidae: Larks

☐ Horned Lark, *Eremophila alpestris*

Family Hirundinidae: Swallows and Martins

☐ Purple Martin, *Progne subis*
☐ Tree Swallow, *Tachycineta bicolor*
☐ Violet-green Swallow, *Tachycineta thalassina*
☐ Northern Rough-winged Swallow, *Stelgidopteryx serripennis*
☐ Bank Swallow, *Riparia riparia*
☐ Cliff Swallow, *Petrochelidon pyrrhonota*
☐ Barn Swallow, *Hirundo rustica*

Family Paridae: Titmice

☐ Black-capped Chickadee, *Poecile atricapillus*
☐ Mountain Chickadee, *Poecile gambeli*
☐ Tufted Titmouse, *Baeolophus bicolor*

Family Sittidae: Nuthatches

☐ Red-breasted Nuthatch, *Sitta canadensis*
☐ White-breasted Nuthatch, *Sitta carolinensis*
☐ Pygmy Nuthatch (*Sitta pygmaea*)

Family Certhiidae: Creepers

☐ Brown Creeper, *Certhia americana*

Family Troglodytidae: Wrens

☐ Rock Wren, *Salpinctes obsoletus*
☐ Carolina Wren, *Thryothorus ludovicianus*
☐ Bewick's Wren, *Thryomanes bewickii*

☐ House Wren, *Troglodytes aedon*
☐ Winter Wren, *Troglodytes troglodytes*
☐ Sedge Wren, *Cistothorus platensis*
☐ Marsh Wren, *Cistothorus palustris*

Family Polioptilidae: Gnatcatchers

☐ Blue-gray Gnatcatcher, *Polioptila caerulea*

Family Regulidae: Kinglets

☐ Golden-crowned Kinglet, *Regulus satrapa*
☐ Ruby-crowned Kinglet, *Regulus calendula*

Family Turdidae: Thrushes and Allies

☐ Eastern Bluebird, *Sialia sialis*
☐ Mountain Bluebird, *Sialia currucoides*
☐ Townsend's Solitaire, *Myadestes townsendi*
☐ Veery, *Catharus fuscescens*
☐ Gray-cheeked Thrush, *Catharus minimus*
☐ Swainson's Thrush, *Catharus ustulatus*
☐ Hermit Thrush, *Catharus guttatus*
☐ Wood Thrush, *Hylocichla mustelina*
☐ American Robin, *Turdus migratorius*
☐ Varied Thrush, *Ixoeus naevius*

Family Mimidae: Mockingbirds, Thrashers, and Catbirds

☐ Gray Catbird, *Dumetella carolinensis*
☐ Northern Mockingbird, *Mimus polyglottos*
☐ Sage Thrasher, *Oreoscoptes montanus*
☐ Brown Thrasher, *Toxostoma rufum*
☐ Curve-billed Thrasher, *Toxostoma curvirostra*
☐ Family Sturnidae: Starlings
☐ European Starling, *Sturnus vulgaris*

Family Motacillidae: Pipits

- ☐ American Pipit, *Anthus rubescens*
- ☐ Sprague's Pipit, *Anthus spragueii*

Family Bombycillidae: Waxwings

- ☐ Bohemian Waxwing, *Bombycilla garrulus*
- ☐ Cedar Waxwing, *Bombycilla cedrorum*

Family Calcariidae: Longspurs and Snow Buntings

- ☐ Lapland Longspur, *Calcarius lapponicus*
- ☐ Chestnut-collared Longspur, *Calcarius ornatus*
- ☐ Smith's Longspur, *Calcarius pictus*
- ☐ McCown's Longspur, *Calcarius mcownii*
- ☐ Snow Bunting, *Plectrophenax nivalis*

Family Parulidae: Wood Warblers

- ☐ Ovenbird, *Seiurus aurocapillus*
- ☐ Worm-eating Warbler, *Helmintheros vermivorum*
- ☐ Louisiana Waterthrush, *Parksia motacilla*
- ☐ Northern Waterthrush, *Parksia noveboracensis*
- ☐ Golden-winged Warbler, *Vermivora chrysoptera*
- ☐ Blue-winged Warbler, *Vermivora pinus*
- ☐ Black-and-White Warbler, *Mniotilta varia*
- ☐ Prothonotary Warbler, *Protonotaria citrea*
- ☐ Swainson's Warbler, *Lymnothlypis swainsonii*
- ☐ Tennessee Warbler, *Oreothlypis peregrina*
- ☐ Orange-crowned Warbler, *Oreothlypis celata*
- ☐ Nashville Warbler, *Oreothlypis ruficapilla*
- ☐ Virginia's Warbler, *Oreothlypis virginiae*
- ☐ Connecticut Warbler, *Oporonis agilis*
- ☐ MacGillivray's Warbler, *Oporornis tolmiei*
- ☐ Mourning Warbler, *Oporonis philadelphia*
- ☐ Kentucky Warbler, *Geothlypis formosa*
- ☐ Common Yellowthroat, *Geothlypis trichas*

☐ American Redstart, *Setophaga ruticilla*
☐ Cape May Warbler, *Setophaga tigrina*
☐ Cerulean Warbler, *Setophaga cerulea*
☐ Northern Parula, *Parula americana*
☐ Magnolia Warbler, *Setophaga magnolia*
☐ Bay-breasted Warbler, *Setophaga castanea*
☐ Blackburnian Warbler, *Setophaga fusca*
☐ Yellow Warbler, *Setophaga petechia*
☐ Chestnut-sided Warbler, *Setophaga pensylvanica*
☐ Blackpoll Warbler, *Setophaga striata*
☐ Black-throated Blue Warbler, *Setophaga caerulescens.*
☐ Palm Warbler, *Setophaga palmarum*
☐ Pine Warbler, *Setophaga pinus*
☐ Yellow-rumped Warbler, *Setophaga coronata*
☐ Yellow-throated Warbler, *Setophaga dominica*
☐ Prairie Warbler, *Setophaga discolor*
☐ [Black-throated Gray Warbler, *Setophaga nigrescens*]
☐ Townsend's Warbler, *Setophaga townsendi*
☐ Black-throated Green Warbler, *Setophaga virens*
☐ Canada Warbler, *Cardellina canadensis*
☐ Wilson's Warbler, *Cardellina pusilla*
☐ Yellow-breasted Chat, *Icteria virens*

Family Emberizidae: Towhees and Sparrows

☐ Spotted Towhee, *Pipilo maculatus*
☐ Eastern Towhee, *Pipilo erythrophthalmus*
☐ American Tree Sparrow, *Spizella arborea*
☐ Chipping Sparrow, *Spizella passerinea*
☐ Clay-colored Sparrow, *Spizella pallida*
☐ Field Sparrow, *Spizella pusilla*
☐ Vesper Sparrow, *Pooecetes gramineus*
☐ Lark Sparrow, *Chondestes grammacus*
☐ Lark Bunting, *Calamospiza melanocorys*
☐ Savannah Sparrow, *Passerculus sandwichensis*

☐ Grasshopper Sparrow, *Ammodramus savannarum*
☐ Baird's Sparrow, *Ammodramus bairdii*
☐ Henslow's Sparrow, *Ammodramus henslowii*
☐ Le Conte's Sparrow, *Ammodramus leconteii*
☐ Nelson's Sparrow, *Ammodramus nelsoni*
☐ Fox Sparrow, *Passerella iliaca*
☐ Song Sparrow, *Melospiza melodia*
☐ Lincoln's Sparrow, *Melospiza lincolnii*
☐ Swamp Sparrow, *Melospiza georgiana*
☐ White-throated Sparrow, *Zonotrichia albicollis*
☐ Harris's Sparrow, *Zonotrichia querula*
☐ White-crowned Sparrow, *Zonotrichia leucophrys*
☐ Dark-eyed Junco, *Junco hyemalis*

Family Cardinalidae: Cardinals, Tanagers and Grosbeaks

☐ Summer Tanager, *Piranga rubra*
☐ Scarlet Tanager, *Piranga olivacea*
☐ Northern Cardinal, *Cardinalis cardinalis*
☐ Rose-breasted Grosbeak, *Pheucticus ludovicianus*
☐ Black-headed Grosbeak, *Pheucticus melanocephalus*
☐ Blue Grosbeak, *Passerina caerulea*
☐ Lazuli Bunting, *Passerina amoena*
☐ Indigo Bunting, *Passerina cyanea*
☐ Painted Bunting, *Passerina ciris*
☐ Dickcissel, *Spiza americana*

Family Icteridae: Blackbirds and Orioles

☐ Bobolink, *Dolichonyx oryzivorus*
☐ Red-winged Blackbird, *Agelaius phoeniceus*
☐ Eastern Meadowlark, *Sturnella magna*
☐ Western Meadowlark, *Sturnella neglecta*
☐ Yellow-headed Blackbird, *Xanthocephalus xanthocephalus*
☐ Rusty Blackbird, *Euphagus carolinus*
☐ Brewer's Blackbird, *Euphagus cyanocephalus*

- ☐ Common Grackle, *Quiscalus quiscula*
- ☐ Great-tailed Grackle, *Quiscalus mexicanus*
- ☐ Brown-headed Cowbird, *Molothrus ater*
- ☐ Orchard Oriole, *Icterus spurius*
- ☐ Bullock's Oriole, *Icterus bullockii*
- ☐ Baltimore Oriole, *Icterus galbula*
- ☐ Scott's Oriole, *Icterus parisorum*

Family Fringillidae: Finches

- ☐ [Gray-crowned Rosy-Finch, *Leucosticte tephrocotis*]
- ☐ Pine Grosbeak, *Pinicola enucleator*
- ☐ Purple Finch, *Carpodacus purpureus*
- ☐ Cassin's Finch, *Carpodacus cassinii*
- ☐ House Finch, *Carpodacus mexicanus*
- ☐ Red Crossbill, *Loxia curvirostra*
- ☐ White-winged Crossbill, *Loxia leucoptera*
- ☐ Common Redpoll, *Acanthus flammea*
- ☐ Pine Siskin, *Spinus pinus*
- ☐ Lesser Goldfinch, *Carduelis psaltria*
- ☐ American Goldfinch, *Carduelis tristis*
- ☐ Evening Grosbeak, *Coccothraustes vespertinus*

Family Fringillidae: Old World Sparrows

- ☐ House Sparrow, *Passer domesticus*

www.ingramcontent.com/pod-product-compliance
Lightning Source LLC
Chambersburg PA
CBHW031203270326
41931CB00006B/388